Lecture Notes in Physics

Edited by H. Araki, Kyoto, J. Ehlers, München, K. Hepp, Zürich
R. Kippenhahn, München, H.A. Weidenmüller, Heidelberg
J. Wess, Karlsruhe and J. Zittartz, Köln
Managing Editor: W. Beiglböck

280

Field Theory, Quantum Gravity and Strings II

Proceedings of a Seminar Series
Held at DAPHE, Observatoire de Meudon,
and LPTHE, Université Pierre et Marie Curie, Paris,
Between October 1985 and October 1986

Edited by H. J. de Vega and N. Sánchez

Springer-Verlag Berlin Heidelberg GmbH

Editors

H. J. de Vega
Université Pierre et Marie Curie, L.P.T.H.E.
Tour 16, 1er étage, 4, place Jussieu, F-75230 Paris Cedex, France

N. Sánchez
Observatoire de Paris, Section d'Astrophysique de Meudon
5, place Jules Janssen, F-92195 Meudon Principal Cedex, France

ISBN 978-3-662-13649-2   ISBN 978-3-540-47934-5 (eBook)
DOI 10.1007/978-3-540-47934-5

© Springer-Verlag Berlin Heidelberg 1987
Originally published by Springer-Verlag Berlin Heidelberg New York 1987
Softcover reprint of the hardcover 1st edition 1987

2153/3140-543210

P R E F A C E

This book contains the lectures delivered in the third year, 1985 - 1986, of the Paris-Meudon Seminar Series.

A seminar series on current developments in mathematical physics was started in the Paris region in October 1983. The seminars are held alternately at the DAPHE-Observatoire de Meudon and LPTHE-Université Pierre et Marie Curie (Paris VI) to encourage theoretical physicists of different disciplines and a number of mathematicians to meet regularly. The seminars delivered in this series in the periods October 1983 - October 1984 and October 1984 - October 1985 have already been published by Springer-Verlag as Lecture Notes in Physics, volumes 226 and 246, respectively.

The present volume "Field Theory, Quantum Gravity and Strings, II" accounts for the lectures delivered up to October 1986. This set of lectures contains selected topics of current interest in field and particle theory, cosmology and statistical mechanics. Basic problems of string and superstring theory are treated in a contemporary perspective and quantum field theoretical as well as string approaches to cosmology are presented. Recent progress on integrable theories and related subjects in two, four and more dimensions is reviewed.

It is a pleasure to thank all the speakers for their successful efforts in delivering comprehensive and stimulating lectures. We thank all the participants for their interest and for their stimulating discussions. We particularly thank the Scientific Direction "Mathématiques - Physique de Base" of C.N.R.S. and the Observatoire de Paris-Meudon for the financial support which made this series possible. We extend our appreciation to Springer-Verlag for their cooperation and efficiency in publishing these proceedings.

CERN, Geneva, February 1987   H. DE VEGA

                                                                N. SANCHEZ

CONTENTS

COVARIANT QUANTIZATION OF THE
BOSONIC STRING: FREE THEORY

P. Di Vecchia

Nordita, Blegdamsvej 17, DK-2100 Copenhagen Ø

The bosonic string is described by the following action[1]

$$S\left[x^\mu(\tau,\sigma)\ ,\ g^{\alpha\beta}(\tau,\sigma)\right] = -\frac{T}{2}\int d\tau \int_0^\pi d\sigma\ \sqrt{-g}\ g^{\alpha\beta}\partial_\alpha x\cdot\partial_\beta x \tag{1}$$

that is classically equivalent to the Nambu-Goto action[2] and is invariant under reparametrizations of the coordinates of the world sheet of the string. A reparametrization induces the following transformations on x^μ and $g_{\alpha\beta}$

$$\delta x^\mu = \epsilon^\alpha\ \partial_\alpha\ x^\mu$$

$$\delta g_{\alpha\beta} = \epsilon^\gamma\ \partial_\gamma\ g_{\alpha\beta} + \partial_\alpha\ \epsilon^\gamma\ g_{\gamma\beta} + \partial_\beta\ \epsilon^\gamma\ g_{\alpha\gamma} \tag{2}$$

where ϵ^α are two arbitrary functions of τ and σ .
The action (1) is in addition also invariant under Weyl transformations:

$$\delta x^\mu = 0 \qquad\qquad \delta g_{\alpha\beta} = 2\Lambda(\tau,\sigma)\ g_{\alpha\beta} \tag{3}$$

where $\Lambda(\tau,\sigma)$ is an arbitrary function of (τ,σ) .
The invariances (2) and (3) are sufficient to gauge away all the components of the metric tensor.
However the Weyl invariance cannot be in general maintained in the quantum theory.
Therefore in the quantization of (1) we can only fix the conformal gauge characterized by the following choice of the metric tensor:

$$g_{\alpha\beta} = \rho(\xi)\ \eta_{\alpha\beta}\ ; \qquad \eta_{11} = -\eta_{00} = 1 \tag{4}$$

where $\rho(\xi)$ is an arbitrary function of $\xi^\alpha \equiv (\tau,\sigma)$.

Since in what follows, however, we will consider only the case of critical dimension $D = 26$, where the Weyl anomaly vanishes[3], we can choose $\rho(\xi) = 1$ in (4).

In the conformal gauge the action (1) becomes[4]:

$$S = -\frac{1}{2\pi} \int d\tau \int_0^\pi d\sigma \left\{ \partial_\alpha x \cdot \partial_\alpha x - 4 \, b_{\alpha\beta} \, \partial^\alpha c^\beta \right\} \tag{5}$$

where the second term is the contribution of the Faddeev-Popov deter-minant obtained from having fixed the conformal gauge. It contains a ghost coordinate c^α and a symmetric and traceless antighost coor-dinate $b^{\alpha\beta}$.

The conformal gauge choice (4) does not fix completely the gauge. We can still perform gauge transformations that leave in the conformal gauge. They are the conformal transformations characterized by two functions $\varepsilon^\alpha(\xi)$ satisfying the condition:

$$\partial^\alpha \varepsilon^\beta + \partial^\beta \varepsilon^\alpha - \eta^{\alpha\beta} \, \partial_\gamma \, \varepsilon^\gamma = 0 \tag{6}$$

In the light-cone coordinates

$$\varepsilon^\pm = \varepsilon^0 \pm \varepsilon^1 \quad ; \quad \xi^\pm = \tau \pm \sigma \quad ; \quad \frac{\partial}{\partial\xi^\pm} = \frac{1}{2}\left(\frac{\partial}{\partial\tau} \pm \frac{\partial}{\partial\sigma}\right) \tag{7}$$

the equations (6) get the simple form:

$$\partial^+ \varepsilon^- = \partial^+ \varepsilon^- = 0 \tag{8}$$

implying that $\varepsilon^+[\varepsilon^-]$ is only a function of $\xi^+[\xi^-]$.

It is easy to check that Lagrangian (5) is invariant under con-formal transformations provided that x_μ, b and c transform as conformal fields with conformal dimension Δ equal to $0,2$ and -1 respectively.

In a conformal invariant theory as the one described by action (5) where the σ-variable varies in a finite domain $(0,\pi)$ it is con-venient to use, instead of $\xi^\pm$, the two variables

$$z = e^{i(\tau+\sigma)} \qquad\qquad \bar{z} = e^{i(\tau-\sigma)} \tag{9}$$

that in euclidean space $(\tau \to i\tau)$ become one the complex conjugate of the other.

A conformal invariant theory is characterized by a conserved and traceless energy-momentum tensor with only two independent components

$T(z)$ and $\bar{T}(\bar{z})$.

A conformal field ϕ with dimension $(\Delta,\bar{\Delta})$ transforms as follows under a conformal transformation:

$$\delta\Phi = \left[\varepsilon(z)\,\frac{\partial}{\partial z} + \Delta\,\varepsilon'(z)\right]\Phi + \left[\bar{\varepsilon}(\bar{z})\frac{\partial}{\partial\bar{z}} + \bar{\Delta}\,\bar{\varepsilon}'(\bar{z})\right]\phi \tag{10}$$

For the sake of simplicity we will omit in the following the dependence on the variable $\bar{z}$, keeping in mind that whatever we do with z can also be done with $\bar{z}$.

The transformation (10) is obtained by requiring the following operator product expansion (OPE) of $T(z)$ with ϕ [5]:

$$T(z)\ \phi(\zeta) = \frac{\partial/\partial\zeta\ \phi(\zeta)}{z-\zeta} + \frac{\phi(\zeta)}{(z-\zeta)^2} + \text{regular terms} \tag{11}$$

The energy-momentum tensor $T(z)$ is a conformal tensor with $\Delta = 2$. This implies [5]:

$$T(z)\ T(\zeta) = \frac{\partial/\partial\zeta\ T(\zeta)}{z-\zeta} + 2\,\frac{T(\zeta)}{(z-\zeta)^2} + \frac{c/2}{(z-\zeta)^4} + \text{reg. terms} \tag{12}$$

An additional more singular c-number term can in general be added without destroying the closure of the conformal algebra. The Virasoro generators can be constructed in terms of $T(z)$:

$$L_n = \oint dz\ z^{n+1}\ T(z) \tag{13}$$

where the integral is defined in such a way that $\oint \frac{dz}{z} = 1$.

From (12) it follows that they satisfy the Virasoro algebra:

$$[L_n\ ,\ L_m] = (n-m)\ L_{n+m} + \frac{c}{12}\ n(n^2-1)\ \delta_{n+m;0} \tag{14}$$

In terms of the variables z and $\bar{z}$ the Lagrangian corresponding to (5) is proportional to:

$$L \sim \partial x\cdot\bar{\partial}x + \frac{1}{2}\,(b\ \bar{\partial}c + \bar{b}\ \partial c) \tag{15}$$

where $\partial \equiv \frac{\partial}{\partial z}$, $\bar{\partial} = \frac{\partial}{\partial\bar{z}}$ and

$$b = b^{zz}\ ;\ \bar{b} = b^{\bar{z}\bar{z}}\ ;\ c = c^z\ ,\ \bar{c} = c^{\bar{z}} \tag{16}$$

Since x_μ , b and c transform as conformal fields with $\Delta = 0,2$ and -1 respectively, it follows that L is a conformal density:

$$\delta L = \partial[\epsilon(z)L] + \bar{\partial}[\bar{\epsilon}(\bar{z})L] \tag{17}$$

implying that the corresponding action is conformal invariant.

The transformations on x_μ , b and c are generated by the following energy-momentum tensor[*]:

$$T(z) = T^x(z) + T^g(z) \tag{18}$$

where

$$T^x(z) = -\frac{1}{2}:\left(\frac{\partial x}{\partial z}\right)^2: \tag{19}$$

$$T^g(z) = : cb' + 2c'b : \tag{20}$$

as it can be seen by using the following contraction rules:

$$<x^\mu(z)\ x^\nu(\zeta)> = -g^{\mu\nu}\ \log(z-\zeta) \tag{21}$$

$$<b(z)\quad c(\zeta)> = \frac{1}{z-\zeta} \tag{22}$$

They allow one to compute also the OPE with two energy-momentum tensors:

$$T(z)\ T(\zeta) = \frac{\partial/\partial\zeta\ T(\zeta)}{z-\zeta} + 2\ \frac{T(\zeta)}{(z-\zeta)^2} + \frac{\frac{D-26}{2}}{(z-\zeta)^4} \tag{23}$$

implying that the c-number of the Virasoro algebra is vanishing at the critical dimension $D = 26$.

As previously explained we have limited our analysis to the quantities that depend on z . In the case of a closed string we can however repeat everything for the quantities that depend on $\bar{z}$ obtaining for instance two sets of mutually commuting Virasoro algebras.

In the case of an open string it is convenient to require that the parametrization of the end points of the string is left unchanged. This implies that $\epsilon(z=e^{i\tau}) = \bar{\epsilon}(z=e^{i\tau})$.

It is easy to convince oneself that for an open string we can

---

[*]In the treatment of the ghost we follow closely the approach of Friedan, Martinec and Shenker[6] .

use all the previous formulas with $z = e^{i\tau}$. In the following we limit for simplicity our considerations to this case.

Having fixed the conformal gauge we have lost the general invariance (2) keeping only the invariance under conformal transformations.

On the other hand we have gained the invariance under BRST transformations, that act as follows on the coordinates of the string:

$$\delta x = \lambda c x'$$

$$\delta b = - 2\lambda x' + \lambda \left[c b' + 2c'b \right] \tag{24}$$

$$\delta c = \lambda c c'$$

where λ is a constant Grassmann parameter.

The variation of Lagrangian (15) under the transformations (24) is a total derivative

$$\delta L = \partial [\lambda cL] \tag{25}$$

implying the invariance of the corresponding action.

It is easy to see that the product of two transformations (24) is identically vanishing.

The generator of the transformations (24) is the BRST charge:

$$Q = \oint dz : c(z) \left[T^x(z) + \frac{1}{2} T^g(z) \right] : \tag{26}$$

By using the contractions (21) and (22) it can be shown after some calculation that:

$$Q^2 = \frac{1}{24} (D-26) \oint d\zeta \ c'''(\zeta) \ c(\zeta) \tag{27}$$

Therefore the quantum BRST charge is nilpotent only if $D = 26$. This implies that our quantization procedure is consistent only for the critical dimension $D = 26$.

In this case the BRST charge commutes with the Virasoro generators:

$$\left[Q \ , \ L_n \right] = 0 \tag{28}$$

for any n .

In conclusion if $D = 26$ the gauge fixed action (5) is invariant under two independent and very important transformations: BRST and conformal transformations.

It is useful to expand $x_\mu(z)$, $b(z)$ and $c(z)$ in terms of the harmonic oscillators. They are given by:

$$x_\mu(z) = q_\mu - ip_\mu \log z + i \sum_{n=1}^{\infty} \frac{1}{\sqrt{n}} \left(a_n z^{-n} - a_n^+ z^n \right)$$

$$c(z) = c_0 z + \sum_{n=1}^{\infty} \left(c_n z^{1-n} + c_n^+ z^{1+n} \right) \tag{29}$$

$$b(z) = \frac{b_0}{z^2} + \sum_{n=1}^{\infty} \left(b_n z^{-2-n} + b_n^+ z^{n-2} \right)$$

The oscillators satisfy the following (anti)-commutation relations:

$$\left[a_{n;\mu} \quad a_{m;\nu}^+ \right] = \delta_{n,m} \, g_{\mu\nu} \qquad [q_\mu, \, p_\nu] = i \, g_{\mu\nu}$$

$$\left\{ c_n \, , \, b_m^+ \right\} = \delta_{n,m} \tag{30}$$

The other (anti)-commutators are vanishing.

The mode expansion for the coordinate $x_\mu(z)$ gives the contraction:

$$<0| \; x_\mu'(z) \; x_\nu'(\zeta) |0> = - \frac{g_{\mu\nu}}{(z-\zeta)^2} \tag{31}$$

Integrating both sides of (31) and setting to zero the two integration constants we get the contraction (21).

In order to derive the contraction (22) from the mode expansion some more discussion is needed.

We can introduce the ghost number current

$$j(z) = :c(z) \; b(z): \tag{32}$$

Using the contraction (22) it is easy to show the following OPE's:

$$j(z) \; j(\zeta) = \frac{1}{(z-\zeta)^2} \tag{33}$$

$$T^g(z) \; j(\zeta) = \frac{\partial/\partial\zeta \; j(\zeta)}{z-\zeta} + \frac{j(\zeta)}{(z-\zeta)^2} - \frac{3}{(z-\zeta)^3} \tag{34}$$

Because of the extra term $j(\zeta)$ is not quite a conformal field with $\Delta = 1$. In terms of the mode expansion defined by

$$j(z) = \sum_n \frac{j_n}{z^{n+1}} \tag{35}$$

(33) and (34) imply:

$$\left[L_n^g , j_m \right] = - m \; j_{n+m} - \frac{3}{2} \; n(n+1) \; \delta_{n+m;0} \tag{36}$$

$$\left[j_m , j_n \right] = n \; \delta_{n+m;0} \tag{37}$$

They can be checked directly using the expansion in harmonic oscillators:

$$j_m = \sum_k : c_{m-k} \, b_k : \tag{38}$$

$$L_n^g = \sum_m (n+m) : b_{n-m} \, c_m : \tag{39}$$

where $c_{-|n|} \equiv c_{|n|}^+$ and $b_{-|n|} \equiv b_{|n|}^+$ and the normal ordering is defined as follows

$$: c_n \, b_{-n} : \begin{cases} = c_n \, b_{-n} & \text{if } n \leq 1 \\[2mm] = - b_{-n} \, c_n & \text{if } n \geq 2 \end{cases} \tag{40}$$

From (38) and (39) one gets that

$$L_1^+ = L_{-1} \qquad\qquad j_{-1} = - j_1^+ \tag{41}$$

and the commutator (36) for $n = \pm 1$ and $m = \pm 1$ implies that the ghost number j_0 is not antihermitian as j_1, but it satisfies the more complicated relation:

$$j_0 + j_0^+ - 3 = 0 \tag{42}$$

If $|q>$ is an eigenstate of the ghost number

$$j_0 |q> = q |q> \tag{43}$$

(42) implies that

$$<q'|q> \sim \delta_{q;3-q'} \tag{44}$$

A state with ghost number q satisfies the relations:

$$b_n|q> = 0 \qquad \text{if} \quad n > q-2$$

$$c_n|q> = 0 \qquad \text{if} \quad n \geq -q+2 \tag{45}$$

that imply

$$L_0|q> = \frac{1}{2} q(q-3)|q> \tag{46}$$

Using (39) and (45) it is possible to show that $|q = 0>$ is the only eigenstate of j_0 that is annihilated by the generators of the projective subgroup of the Virasoro algebra:

$$L_0|q = 0> = L_1|q = 0> = L_1^+|q = 0> = 0 \tag{47}$$

$|q = 0>$ is therefore projective invariant.

After these considerations it is clear that, because of (44), we cannot get a non vanishing result in (22) unless we use a bra and a ket state, whose ghost number differs by 3 . In particular in order to get (22) we must compute:

$$<q = 3| \quad b(z) \quad c(\zeta) \quad |q = 0> \tag{48}$$

as it can be simply shown by using (45).

Using the mode expansion we can compute the BRST charge Q in terms of the oscillators. It is given by:

$$Q = \sum_{n=1}^{\infty} \left[c_n L_{-n}^x + c_n^+ L_n^x \right] + c_0 \left[L_0^x + L_0^g \right] + \tilde{Q} \tag{49}$$

where

$$\tilde{Q} = \sum_{n,m=1}^{\infty} m \left[c_n^+ c_m^+ b_{n+m} - c_n c_m b_{n+m}^+ \right] - 2b_0 \sum_1^{\infty} n c_n^+ c_n$$

$$- \sum_{n,m=1}^{\infty} (n+2m) \left[c_m^+ c_{n+m} b_n^+ + c_{n+m}^+ c_m b_n \right] \tag{50}$$

From (49) it follows that the state $|q = 0>$ is also BRST invariant:

$$Q|q = 0> = 0 \tag{51}$$

In the BRST quantization one treats all Lorentz components on equal ground keeping the manifest Lorentz invariance of the theory. However the space, in which the system is quantized, contains states with negative norm. In order to construct a consistent quantum theory we must require that the physical states span a positive definite subspace. Its elements are characterized by the vanishing of the BRST charge:

$$Q \, |Phys\rangle \; = 0 \tag{52}$$

Because of the nilpotency of Q any state of the form $|\psi\rangle + Q|\lambda\rangle$ is a solution of (51) if the state $|\psi\rangle$ itself satisfies (52) and $|\lambda\rangle$ is arbitrary. In other words the physical states are identified with the cohomology classes of Q.

If we restrict ourselves to states with no ghost excitation of the type $|q = 1\rangle_{b,c} \otimes |\psi\rangle_a$ it is easy to see that (51) reduces to the well known conditions on the physical states[7]:

$$L_n|\psi\rangle \; = (L_0-1) \; |\psi\rangle \; = 0 \tag{53}$$

as it follows directly from (49).

The state $|q = 1\rangle$ is given in terms of the state $|q = 0\rangle$ by

$$|q = 1\rangle = c_1|q = 0\rangle \tag{54}$$

and it is annihilated by all annihilation oscillators and by b_0, but not by c_0.

It is important to notice that, unless the orbital modes, the state (54) used to construct physical states is not annihilated by the generators of the projective subalgebra.

The most general solution of eqs. (53) for $D = 26$ is provides by the transverse states[8] defined in terms of the following operator:

$$A_{i;n} = \oint dz \; x'_\mu \cdot \varepsilon_i^\mu \; e^{ik \cdot x(z)} \tag{55}$$

where the index i runs over the 24 transverse directions, that are orthogonal to k_μ. Because of the $\log z$ appearing in $e^{ik \cdot x(z)}$, as one can see in (29), the integral in (55), that is performed around the origin $z = 0$, is well defined only if we constraint the momentum of the state, on which $A_{i,n}$ acts, to satisfy the relation

$$p \cdot k = - n \tag{56}$$

with an integer n.

Two very important properties of the transverse operator (55) are the following. They commute with the operators L_m:

$$\left[L_m , A_{n;i}\right] = 0 \tag{57}$$

for any integer m and they satisfy the algebra of a non relativistic harmonic oscillator[9]

$$\left[A_{n,i} , A_{m,j}\right] = n \, \delta_{ij} \, \delta_{n+m;0} \tag{58}$$

as it can be shown by using the contraction (31).

In terms of (55) we can construct a complete and orthogonal basis in the space of physical states, that is given by:

$$\prod_n \frac{\left[A_{i_n; -N_n}\right]^{\lambda_n}}{\sqrt{N_n}} |0,p> \tag{59}$$

where $N_n > 0$.

The states (59) satisfy the physical conditions (53) as it follows from (57) and span a positive definite space. This implies that the subspace of physical states is ghost free[10].

ACKNOWLEDGEMENTS

I wish to thank J. L. Petersen for many useful discussions on BRST quantization.

REFERENCES

1) L. Brink, P. Di Vecchia and P. Howe, Phys. Lett. 65B (1976) 471
   S. Deser and B. Zumino, Phys. Lett. 65B (1976) 369
2) Y. Nambu, Lectures at the Copenhagen Symposium, 1970, unpublished
   T. Goto, Progr. Theor. Phys. 46 (1971) 1560
3) A. M. Polyakov, Phys. Lett. 103B (1981) 502
4) D. Friedan, "Introduction to Polyakov's String Theory" in Recent Advanced in Field Theory and Statistical Mechanics (Les Houches 1982)
   M. Kato and K. Ogawa, Nucl. Phys. B212 (1983) 443
   S. Hwang, Phys. Rev. D28 (1983) 2614
5) A. A. Belavin, A. M. Polyakov and A. B. Zamolodchikov, Nucl. Phys. B241 (1984) 333
6) D. Friedan, E. Martinec and S. Shenker, Nucl. Phys. B271 (1986) 93
7) E. Del Giudice and P. Di Vecchia, Nuovo Cimento 70A (1970) 579
8) E. Del Giudice, P. Di Vecchia and S. Fubini, Annals of Physics 70 (1972) 378
9) R. Brower and P. Goddard, Nucl. Phys. B40 (1972) 437
10) R. C. Brower, Phys. Rev. D6 (1972) 1655
    P. Goddard and C. B. Thorn, Phys. Lett. 40B (1972) 235

OPERATORIAL QUANTIZATION OF DYNAMICAL SYSTEMS WITH IRREDUCIBLE FIRST AND SECOND CLASS CONSTRAINTS

I.A. Batalin and E.S. Fradkin
Lebedev Physical Institute, Moscow

Abstract

Operatorial version is suggested of the generalized canonical quantization method of dynamical systems subjected to irreducible first and second class constraints. An operatorial analog of classical Dirac brackets is realized. Generating equations for generalized algebra of first and second class constraints, as well as for the unitarizing Hamiltonian are formulated. In the first class constraint sector new generating equations are presented directly in terms of operatorial Dirac brackets.

Introduction

During recent years a method of generalized canonical quantization of constrained dynamical systems has been being developed in the works of the group of authors [1-10]. The cornerstone of the method is the idea [1] that constrained systems admit canonical commutation relations in an extended phase space which includes, along with the initial variables, also dynamically active Lagrange multipliers and ghosts. The physical unitarity and gauge independence are provided within this approach via dynamical compensation of the contributions of Lagrange multipliers and ghosts, for which possibility their opposite statistics is responsible. Until recently this idea was directly applied as a matter of fact only to the first class constraints. The second class constraints were handled by using canonical measure on the corresponding hypersurface [11] in the path integral and the Dirac brackets in the generating equations of the gauge algebra [4,5]. The lack of a relevant formal scheme that would admit the use of canonical commutation relations in the case when second class constraints are present too, was a serious obstacle in realizing the program of operatorial quantization in the most general case. In our previous work [12] this obstacle was overcome, and an operatorial version of the method of generalized canonical quantization of dynamical systems subject to second class constraints was formulated. The goal of the present paper is to include a more general case into the framework of the work [12] when first class constraints are also initially present. In the present context both first and second class constraints are assumed to be linearly independent (irreducible).

<u>Designations.</u> The same as in our previous works $\varepsilon(A)$ designates the Grassmann parity of the quantity A. The supercommutator of operators A and B is defined as

$$[A,B] \equiv AB - BA(-1)^{\varepsilon(A)\varepsilon(B)} . \tag{0.1}$$

We write every canonical pair (momentum and co-ordinate) as

$$(P_A, Q^A), \; \varepsilon(P_A) = \varepsilon(Q^A), \; A = 1, \ldots, N, \tag{0.2}$$

so that the only nonzero equal-time supercommutators for them are

$$[Q^A, P_B] = i\hbar \delta_B^A . \tag{0.3}$$

1. Generating Equations of Generalized Algebra of Constraints

Let

$$(p_i, q^i), \; \varepsilon(p_i) = \varepsilon(q^i), \; i = 1, \ldots, n \tag{1.1}$$

be initial pairs of canonically conjugate operators. Let a dynamical system be given in the phase space (1.1) with the Hamiltonian

$$H_0 = H_0(p,q), \; \varepsilon(H_0) = 0 \quad , \tag{1.2}$$

irreducible first class constraints

$$T'_a = T'_a(p,q), \; \varepsilon(T'_a) \equiv \varepsilon'_a, \; a = 1, \ldots, m', \tag{1.3}$$

and irreducible second class constraints

$$T''_\alpha = T''_\alpha(p,q), \; \varepsilon(T''_\alpha) \equiv \varepsilon''_\alpha, \; \alpha = 1, \ldots, 2m''. \tag{1.4}$$

Consider a pair of canonically conjugate ghost operators for each constraint (1.3), (1.4), whose statistics is opposite to that of the corresponding constraint

$$(\bar{\mathcal{P}}'_a, C'^a), \; \varepsilon(\bar{\mathcal{P}}'_a) = \varepsilon(C'^a) = \varepsilon'_a + 1, \; a = 1, \ldots, m', \tag{1.5}$$

$$(\bar{\mathcal{P}}''_\alpha, C''^\alpha), \; \varepsilon(\bar{\mathcal{P}}''_\alpha) = \varepsilon(C''^\alpha) = \varepsilon''_\alpha + 1, \; \alpha = 1, \ldots, 2m''. \tag{1.6}$$

Initial canonical pairs (1.1) form together with the canonical ghost pairs (1.5), (1.6), the so-called minimal sector. Let us attribute some inner characteristic values to these operators, called the ghost numbers. Consider two classes of the ghost numbers, (gh') and (gh") following the division of the full set of constraints into those of first and second class:

$$gh'(q) = -gh'(p) = 0, \quad gh''(q) = -gh''(p) = 0, \tag{1.7}$$

$$gh'(C') = -gh'(\mathcal{P}') = 1, \quad gh''(C') = -gh''(\mathcal{P}') = 0, \tag{1.8}$$

$$gh'(C'') = -gh'(\mathcal{P}'') = 0, \quad gh''(C'') = -gh''(\mathcal{P}'') = 1. \tag{1.9}$$

By definition, we have for every operator having a ghost number

$$gh'(AB) = gh'(A) + gh'(B), \quad gh''(AB) = gh''(A) + gh''(B). \tag{1.10}$$

Consider the following operatorial equations in the minimal sector[1] (1.1), (1.5), (1.6)

$$[\Omega''_{,}\Omega''] = i\hbar \Omega''^{\alpha} \omega_{\alpha\beta} \Omega''^{\beta}, \tag{1.11}$$

$$[\Omega''^{\alpha}, \Omega''] = 0, \quad [\Omega''^{\alpha}, \Omega''^{\beta}] = 0, \tag{1.12}$$

$$\epsilon(\Omega'') = 1, \quad gh'(\Omega'') = 0, \quad gh''(\Omega'') = 1, \tag{1.13}$$

$$\epsilon(\Omega''^{\alpha}) = \epsilon''_{\alpha} + 1, \quad gh'(\Omega''^{\alpha}) = 0, \quad gh''(\Omega''^{\alpha}) = 1, \tag{1.14}$$

where $\omega_{\alpha\beta}$ is a c-numerical inversible matrix, such that

$$\epsilon(\omega_{\alpha\beta}) = \epsilon''_{\alpha} + \epsilon''_{\beta}, \quad \omega_{\beta\alpha} = -\omega_{\alpha\beta}(-1)^{\epsilon''_{\alpha}\epsilon''_{\beta}} . \tag{1.15}$$

Solution of equations (1.11-1.14) for operators Ω'', Ω''^{α} is looked for in the form of $\mathcal{P}$C-normal ordered (i.e. with every $\mathcal{P}'$, $\mathcal{P}''$ placed to the left of every C',C")[2] series in powers of the ghost operators (1.5), (1.6), the first term in the

---

[1] For the sake of universality and generality we admit here that the generating operators Ω'', Ω''^{α} of the algebra of second class constraints may depend on the ghosts (1.5) of the first class constraint sector. Note, however, that there always exists a solution of the generating equations (1.11-1.14) which does not depend on operator (1.5) and is quite sufficient for us.

---

[2] The same as in our previous works on operatorial quantization we are using the $\mathcal{P}$C-normal form for the ghost operators. Certainly, we might exploit instead any other normal ordering, e.g. C$\mathcal{P}$-ordering, or the Weyl ordering, since all the normal orderings may be related to one another using the canonical commutation relations.

expansion of the operator Ω'' being $T_\alpha'' C_\alpha''$. Substituting the $\mathcal{P}C$-expansions of the operators Ω'', Ω''^α into equations (1.11), (1.12), and reducing their left- and right-hand sides to $\mathcal{P}C$-normal form we obtain a sequence of relations for the coefficient operators to be solved step by step. In this way structural relations of the generalized algebra of second class constraints are generated within the generating equations (1.11-1.14).

Consider now how the gauge algebra of first class constraints is generated. To this end introduce, first of all, operators $\bar{\Omega}''_\alpha$, canonically conjugate to Ω''^α:

$$[\Omega''^\alpha, \bar{\Omega}''_\beta] = i\hbar \delta^\alpha_\beta, \quad [\bar{\Omega}''_\alpha, \bar{\Omega}''_\beta] = 0, \quad \bar{\Omega}''_\alpha \Omega''^\alpha = \mathcal{P}''_\alpha C''^\alpha, \tag{1.16}$$

$$\epsilon(\bar{\Omega}''_\alpha) = \epsilon''_\alpha + 1, \quad gh'(\bar{\Omega}'') = 0, \quad gh''(\bar{\Omega}'') = -1. \tag{1.17}$$

To each operator A we may put into correspondence the solution $\tilde{A}(\varphi)$ of the following problem with the operator A as an initial datum:

$$\frac{\partial_r \tilde{A}}{\partial \varphi^\alpha} = (i\hbar)^{-1} [\tilde{A}, (i\hbar)^{-1}[\Omega'', \bar{\Omega}''_\alpha]], \quad \tilde{A}(\varphi=0) = A, \tag{1.18}$$

where

$$\varphi^\alpha, \quad \epsilon(\varphi^\alpha) = \epsilon''_\alpha, \quad gh'(\varphi) = gh''(\varphi) = 0, \quad \alpha = 1, \ldots, 2m'' \tag{1.19}$$

are c-numerical parameters. The formal integrability conditions for the problem (1.18) are fulfilled due to the generating equations (1.11-1.14), (1.16), (1.17).

Operatorial Dirac bracket of any two operators A and B is defined as follows

$$[A,B]_{\mathcal{D}} \equiv (\tilde{A}(\varphi_1)\tilde{B}(\varphi_2) - \tilde{B}(\varphi_1)\tilde{A}(\varphi_2)(-1)^{\epsilon(A)\epsilon(B)}$$

$$\times \left. \exp\left\{\frac{i\hbar}{2} \frac{\overleftarrow{\partial}_r}{\partial\varphi^\alpha_1} \omega^{\alpha\beta} \frac{\overrightarrow{\partial}_r}{\partial\varphi^\beta_2}\right\}\right|_{\varphi_1 = \varphi_2 = 0}, \tag{1.20}$$

where $\omega^{\alpha\beta}$ is the matrix inverse to the matrix $\omega_{\alpha\beta}$ from (1.11), (1.15):

$$\omega_{\alpha\beta}\omega^{\beta\gamma} = \delta^\gamma_\alpha, \quad \omega^{\beta\alpha} = \omega^{\alpha\beta}(-1)^{(\epsilon''_\alpha+1)(\epsilon''_\beta+1)}. \tag{1.21}$$

One can show that the Dirac bracket (1.20) possesses every algebraic property of supercommutator defined as (0.1).

Using the definition (1.20) the generating equations of the gauge algebra of the first class constraints may be written as

$$[\Omega', \Omega']_{\mathcal{D}} = 0, \quad [\Omega''^\alpha, \Omega'] = 0, \quad [\Omega', \bar{\Omega}''_\alpha] = 0, \tag{1.22}$$

$$[H',\Omega']_{\mathscr{D}} = 0, \quad [\Omega''^{\alpha},H'] = 0, \quad [H',\bar{\Omega}''_{\alpha}] = 0, \tag{1.23}$$

$$\varepsilon(\Omega') = 1, \quad gh'(\Omega') = 1, \quad gh''(\Omega') = 0, \tag{1.24}$$

$$\varepsilon(H') = 0, \quad gh'(H') = 0, \quad gh''(H') = 0. \tag{1.25}$$

Solution of these equations for the operators Ω' and H' is looked for in the form of $\mathscr{P}C$-normal-ordered series expansions in powers of the ghosts (1.5), (1.6), the first terms of the $\mathscr{P}C$-expansions for Ω' and H' being $T'_a C'^a$ and H_0 respectively. Substituting these expansions into equations (1.22), (1.23) and reducing their l.-h. sides to the $\mathscr{P}C$-normal form, one obtains a sequence of recurrency relations for finding the coefficient operators. These relations are the structural relations for the gauge algebra of the first class constraints.

2. Unitarizing Hamiltonian

We proceed here by introducing new operators. Consider first the operators

$$\Phi^{\alpha}, \quad \varepsilon(\Phi^{\alpha}) = \varepsilon''_{\alpha}, \quad gh'(\Phi) = gh''(\Phi) = 0, \quad \alpha = 1, \ldots, 2m'', \tag{2.1}$$

which obey the equal-time commutation relations

$$[\Phi^{\alpha},\Phi^{\beta}] = i\hbar\omega^{\alpha\beta} (-1)^{\varepsilon''_{\beta}}, \tag{2.2}$$

(see also (1.21)) and commute with every operator (1.1), (1.5), (1.6) as well as with every operator to be introduced in what follows. Second, extend the sectors (1.5), (1.6) by considering new canonical pairs. In addition to the first class ghosts (1.5), let us introduce the following new canonically conjugate operator pairs

$$(\pi'_a,\lambda'^a), \quad \varepsilon(\pi'_a) = \varepsilon(\lambda'^a) = \varepsilon'_a, \quad a = 1, \ldots, m', \tag{2.3}$$

$$(\bar{C}'_a,\mathscr{P}'^a), \quad \varepsilon(\bar{C}'_a) = \varepsilon(\mathscr{P}'^a) = \varepsilon'_a + 1, \quad a = 1, \ldots, m', \tag{2.4}$$

with the ghost numbers fixed as follows

$$gh'(\lambda') = -gh'(\pi') = 0, \quad gh''(\lambda') = -gh''(\pi') = 0, \tag{2.5}$$

$$gh'(\mathscr{P}') = -gh'(\bar{C}') = 1, \quad gh''(\mathscr{P}') = -gh''(\bar{C}') = 0. \tag{2.6}$$

Analogously, in addition to the second class ghosts (1.6) let us consider new canonical pairs

$$(\pi'',\lambda''^\alpha), \quad \varepsilon(\pi''_\alpha) = \varepsilon(\lambda''^\alpha) = \varepsilon''_\alpha, \quad \alpha = 1, \ldots, 2m'', \tag{2.7}$$

$$(\overline{C}''_\alpha,\mathcal{P}''^\alpha), \quad \varepsilon(\overline{C}''_\alpha) = \varepsilon(\mathcal{P}''^\alpha) = \varepsilon''_\alpha + 1, \quad \alpha = 1, \ldots, 2m'', \tag{2.8}$$

with the ghost numbers

$$gh'(\lambda'') = -gh'(\pi'') = 0, \quad gh''(\lambda'') = -gh''(\pi'') = 0, \tag{2.9}$$

$$gh'(\mathcal{P}'') = -gh'(\overline{C}'') = 0, \quad gh''(\mathcal{P}'') = -gh''(C'') = 1. \tag{2.10}$$

Let $\widetilde{A}(\varphi)$ be the solution of the problem (1.18), put into correspondence to every operator A, taken as an initial datum. We shall need the following designation

$$:\widetilde{A}(\Phi): \equiv \widetilde{A}(\varphi)\exp\left\{\frac{\overleftarrow{\delta}_r}{\delta\varphi^\alpha}\Phi^\alpha\right\}\Bigg|_{\varphi = 0}, \tag{2.11}$$

where Φ^α are operators from (2.1), (2.2). With this designation define the Fermion operator

$$\Omega \equiv :\widetilde{\Omega}'(\Phi): + \pi^1_a\mathcal{P}'^a + \Omega'' + \omega_{\alpha\beta}\Phi^\beta\Omega''^\alpha + \pi''_\alpha\mathcal{P}''^\alpha. \tag{2.12}$$

Due to (1.11-1.14), (1.16-1.18), (1.22), (1.24) the operator (2.12) is nilpotent:

$$[\Omega,\Omega] = 0. \tag{2.13}$$

Consider next the initial gauge first class Fermion depending on the canonical pairs (1.1), (1.5), (2.3), (2.4):

$$\Psi'_0 = \mathcal{P}'_a\lambda'^a + \overline{C}'_a\chi'^a, \tag{2.14}$$

where

$$\chi'^a, \quad \varepsilon(\chi'^a) = \varepsilon^1_a, \quad gh'(\chi') = gh''(\chi') = 0, \quad a = 1, \ldots, m', \tag{2.15}$$

are operators that fix an admissible gauge in the first class constraint sector. Define a modified (Dirac) gauge Fermion Ψ' using the equations

$$[\Omega''^\alpha,\Psi'] = 0, \quad [\Psi',\overline{\Omega}''_\alpha] = 0, \quad gh'(\Psi') = -1, \quad gh''(\Psi') = 0 \tag{2.16}$$

to be solved by a $\mathcal{P}C$-normal-ordered series in powers of ghosts with (2.14) as the first term. With the solution of equations (2.16) at our disposal we may define the full gauge Fermion

$$\Psi = \,:\widetilde{\Psi}{}'(\Phi):\, + \Psi'', \tag{2.17}$$

where

$$\Psi'' = \overline{\mathcal{P}}''_\alpha \lambda''^\alpha + \overline{C}''_\alpha \chi''^\alpha \tag{2.18}$$

is the second class gauge Fermion, depending on the canonical pairs (1.1), (1.6), (2.1), (2.7), (2.8), while

$$\chi''^\alpha, \quad \varepsilon(\chi''^\alpha) = \varepsilon''_\alpha, \quad gh'(\chi'') = gh''(\chi'') = 0, \quad \alpha = 1, \ldots, 2m'' \tag{2.19}$$

are operators that fix admissible gauge in the second class constraint sector. The full unitarizing Hamiltonian of the theory is given as [13]

$$H = \,:\widetilde{H}{}'(\Phi):\, + (i\hbar)^{-1}[\Psi, \Omega]. \tag{2.20}$$

Operator (2.12) is conserved owing to (1.11-1.14), (1.16-1.18), (1.23), (1.25), (2.13):

$$[H, \Omega] = 0. \tag{2.21}$$

Physical states of the theory are selected by the condition

$$\Omega|\text{Phys}\rangle = 0, \quad |\text{Phys}\rangle \neq \Omega|\text{\scriptsize} \iota\iota\iota\rangle, \tag{2.22}$$

where $|\iota\iota\iota\rangle$ stands for any state.

The physical S-matrix induced by the Hamiltonian (2.20) does not depend on any special choice of admissible gauge operators (2.15), (2.19) and is unitary in the subspace (2.22).

References

1. E.S. Fradkin, G.A. Vilkovisky: Phys. Lett. **55B** (1975) 224
2. E.S. Fradkin, G.A. Vilkovisky, CERN Report TH-2332 (1977)
3. I.A. Batalin, G.A. Vilkovisky: Phys. Lett. **69B** (1977) 309
4. E.S. Fradkin, T.E. Fradkina: Phys. Lett. **72B** (1978) 343
5. I.A. Batalin, E.S. Fradkin: Phys. Lett. **122B** (1983) 157
6. I.A. Batalin, E.S. Fradkin: Phys. Lett. **128B** (1983) 303
7. I.A. Batalin, E.S. Fradkin: J. Math. Phys. **25** (1984) 2426
8. I.A. Batalin, E.S. Fradkin: J. Nucl. Phys. (USSR) **39** (1984) 23
9. I.A. Batalin: J. Nucl. Phys. (USSR) **41** (1985) 278
10. I.A. Batalin, E.S. Fradkin: Rivista Nuovo Cimento (1986) [in press]
11. E.S. Fradkin: Acta Universitatis Wratislaviensis N 207. Proc. Xth Winter School of Theoretical Physics Karpacz (1973) p.93
12. I.A. Batalin, E.S. Fradkin: Preprint P.N. Lebedev Inst. (1986) N 132, Phys. Lett. (1986) [in press]

13. Generally, we might have used any gauge Fermion in (2.20) depending on the complete set of dynamical variables, under the only requirement that it should produce admissible gauge conditions in the first and second class constraint sectors. We have preferred here, however, a somewhat more special type of the gauge Fermion, namely the one given as (2.14-2.19), pursuading the fulfilment of a natural requirement that lifting the gauge degeneracy in the sectors of first and second class constraints should occur in independent ways. Actually this independence is contained in the following two properties. Firstly, we have

$$[:\widetilde{\Psi}'(\Phi):, \ :\widetilde{\Omega}'(\Phi):] = \overbrace{:[\Psi',\Omega]}^{}{}_{\mathcal{D}}:$$

and, secondly, the operator $:\widetilde{\Psi}'(\Phi):$ evidently commutes with the part of operator (2.12), which is marked by two primes indicating the fact that it concerns the second class constraint sector.

KALUZA-KLEIN APPROACH TO SUPERSTRINGS

M.J. Duff

Theory Division, CERN, 1211 Geneva 23, Switzerland

ABSTRACT

We apply Kaluza-Klein techniques to the bosonic string compactified on the $E_8 \times E_8$ group manifold to derive properties of ten-dimensional superstrings, thus lending support to the idea that the bosonic string is the fundamental theory. We then pose the question of why physical space-time has just four dimensions.

1. FERMIONS FROM BOSONS

The appearance of the rank 16, dimension 496, gauge groups $E_8 \times E_8$ and spin $32/Z_2$ as the only available candidates for anomaly free ten-dimensional superstrings[1] prompted Freund[2] to conjecture that the fundamental theory might be the 26-dimensional bosonic string, and that the ten-dimensional theories emerge after compactification on the torus $T^{16} = R^{16}/\Gamma$ where Γ is the even self-dual Euclidean lattice of $E_8 \times E_8$ or spin $32/Z_2$. In this picture, the fermions would appear as solitons of the bosonic theory. In addition to the 16 $[U(1)]^{16}$ elementary Kaluza-Klein gauge bosons, a further 480 gauge bosons would also emerge as Frenkel-Kac[3] solitons since the string can wrap around the torus. The subsequent discovery of the $E_8 \times E_8$ and spin $32/Z_2$ heterotic strings[4] brought the total number consistent superstrings to 5 as in Table 1, and only increased the desire for one underlying theory.

Table 1
Consistent superstrings

TYPE	SPINOR	STRING	LOW ENERGY THEORY
I [SO(32)]	Weyl + Majorana	open/closed	N = 1 supergravity + SO(32) Yang-Mills
IIA	Majorana	closed	N = 2 non-chiral supergravity
IIB	Weyl	closed	N = 2 chiral supergravity
Heterotic [SO(32)]	Weyl + Majorana	closed	N = 1 supergravity + SO(32) Yang-Mills
Heterotic [$E_8 \times E_8$]	Weyl + Majorana	closed	N = 1 supergravity + $E_8 \times E_8$ Yang-Mills

Noting that the bosonic string can also undergo spontaneous compactification from D to d dimensions on the simply-laced non-Abelian group manifold G of radius R provided[5] R = $\sqrt{\alpha'}$ and

$$D - 26 = \dim G \cdot \frac{c_A}{2 + c_A} = \dim G - \text{rank } G \qquad (1)$$

where c_A is the second order Casimir in the adjoint representation, Nilsson, Pope and myself[6] proposed obtaining the d = 10 heterotic strings by choosing G = $E_8 \times E_8$ or spin $32/Z_2$ for which C_A = 60 and hence D = 506, d = 10. The origin of fermions was (and still is) less obvious than in the T^{16} compactification since G is simply connected. However, the appearance of the $E_8 \times E_8$ or spin $32/Z_2$ gauge bosons is easier to understand than the T^{16} case, since they are all just _elementary_ Kaluza-Klein[7] fields, i.e., the gauge groups are just subgroups of the D = 506 general co-ordinate group. One nice feature of this group manifold approach is thus to maintain the Kaluza-Klein ideal of getting internal symmetries from space-time symmetries in a higher dimension; a feature which the superstrings with their primary Yang-Mills fields seemed to lack. The main reason why this traditional Kaluza-Klein idea fell out of favour was its inability to explain chiral fermions. This problem is now avoided by the bosonic string; we simply cut the Gordian knot and dispense with fermions altogether! Moreover, the old Kaluza-Klein trick of adding a cosmological constant in the higher dimensional theory with sign and magnitude so designed to cancel the one arising from compactification is now more respectable, since the required D-dimensional cosmological term $(D-26)/\alpha'$ is enforced by conformal invariance of the string. The Kaluza-Klein relation $R^2 g^2 = \kappa^2$ between the Yang-Mills coupling g and the gravitational constant κ meant that the heterotic strings for which $\alpha' g^2 = \kappa^2$ might indeed admit of such an interpretation, but the Type I string for which $g^2 = \kappa\alpha'$ would not.

Independently, at about the same time, Casher et al.[8] took the idea one step further and showed how <u>all</u> closed superstrings; Type IIA, Type IIB, heterotic $E_8 \times E_8$ and heterotic spin $32/Z_2$ could emerge from T^{16} compactification of the D = 26 bosonic string. The key idea was the identification of the transverse space-time SO(8) of the superstring with the diagonal subgroup of the transverse space-time SO(8) of the compactified bosonic string and SO(8) [internal]:

$$SO(8) \left[\text{space-time, super} \right] \subset$$
$$SO(8) \left[\text{space-time, bosonic} \right] \times SO(8) \left[\text{internal} \right] \tag{2}$$

where SO(8) [internal] is a subgroup of G_R in the heterotic case and $G_L \times G_R$ in the case of Type II. In this way states transforming as spinor representations of SO(8) [internal] now transform as <u>fermion</u> representations of SO(8) [space-time,super].

Since the non-linear σ-model on the group manifold is equivalent to free bosons on the torus[5], it follows that our Kaluza-Klein approach and the Frenkel-Kac approach of Casher et al. are in fact equivalent. Just like the wave-particle duality of quantum mechanics which picture one chooses is merely a matter of convenience. So far, the torus approach has proved more powerful for formal "stringy" results, whereas the elementary nature of the gauge fields in the Kaluza-Klein approach lends itself more readily to low-energy field theory considerations. A striking example of this is the derivation of the d = 10 Lorentz and Yang-Mills Chern-Simons terms summarized in Table 2. The identification of spin-connections with gauge potentials[9] is, as discussed by Nilsson,

Pope, Warner and myself[10] just the field theoretic realization of the diagonal choice of space-time SO(8) discussed above. The identification of A with ω_- in going from the heterotic string to the Type II string had already been employed in the literature[11]. What is not generally appreciated, however, is that going from heterotic (one gravitino) to Type II (two gravitinos) requires exactly the same "fermions from bosons" phenomenon as going from bosonic (zero gravitinos) to heterotic (one gravitino). It is strange, therefore, that physicists who feel at ease with the first still remain sceptical about the second.

Table 2

Kaluza-Klein origin of d = 10 Chern-Simons terms. $\omega_\pm$ are the spin-connections with torsion $\pm\frac{1}{2}H$ and $(A,\tilde{A})$ are the Yang-Mills gauge potentials of (G_L, G_R).

STRING	DIMENSION	CONNECTIONS	(CURVATURE)^2 TERMS	CHERN-SIMONS TERMS
bosonic	506	ω_+, ω_-	$R_+^2 + R_-^2$	$dH = 0$
bosonic on G	10	$\omega_+, \tilde{A}, \omega_-, A$	$R_+^2 - \tilde{F}^2 + R_-^2 - F^2$	$dH = \alpha' \mathrm{Tr}(\tilde{F}\ \tilde{F} - F\ F)$
heterotic	10	$\omega_+ = \tilde{A}, \omega_-, A$	$R_-^2 - F^2$	$dH = \alpha' \mathrm{Tr}(R_+\ R_+ - F\ F)$
Type II	10	$\omega_+ = \tilde{A}, \omega_- = A$	0	$dH = \alpha' \mathrm{Tr}(R_+\ R_+ - R_-\ R_-)$

We shall now outline in Sections 2 and 3 how the derivation goes omitting the details. A more thorough discussion, including a review of the whole Kaluza-Klein programme and its merger with string theory can be found in Refs. 12) and 13). Finally, in Section 4, we pose the question: if the bosonic string really is fundamental theory does this explain why the number of uncompactified space-time dimensions is just four?

2. THE BOSONIC STRING ON THE GROUP MANIFOLD

Our starting point is the background field Lagrangian

$$L = \sqrt{\gamma}\, \gamma^{ab} \partial_a x^M \partial_b x^N \hat{g}_{MN}(x) + \varepsilon^{ab} \partial_a x^M \partial_b x^N \hat{b}_{MN}(x) \qquad (3)$$
$$+ \alpha' \sqrt{\gamma}\, R(\gamma)\, \hat{\phi}(x) + \cdots$$

where $x^M(\xi)$ defines the embedding of the two-dimensional string worldsheet M_2 in a space-time M_D (M,N = 1,...,D), $\xi^a = (\tau,\sigma)$ are the co-ordinates on M_2 and $R(\gamma)$ is the curvature scalar of the worldsheet metric γ_{ab}. The graviton $\hat{g}_{MN}(x)$, the antisymmetric tensor $\hat{B}_{MN}(x)$ and the dilaton $\hat{\phi}(x)$ correspond to the massless models of the bosonic string spectrum. The dots refer to terms describing the higher spin massive modes and the scalar "tachyon". By ignoring these higher modes in (3), we are implicitly assuming that in the correct vacuum state of the theory, these fields have vanishing vevs. (See, however, the cautionary remarks[13] about a possible "space-invaders" phenomenon.)

For consistency, the two-dimensional theory must be conformally invariant and hence the two-dimensional worldsheet stress tensor must be traceless, i.e., there must be no conformal anomaly. One can show that the absence of trace anomaly places restrictions on the background fields $\hat{g}_{MN}(x)$, $\hat{B}_{MN}(x)$ and $\hat{\phi}(x)$ which are equivalent to the Einstein-matter field equations obtained from the effective Lagrangian

$$\mathcal{L}_D = \sqrt{-\hat{g}}\, e^{-2\hat{\phi}} \left[\frac{D-26}{3\alpha'} - \hat{R} - 4(\partial\hat{\phi})^2 + \frac{1}{12}\hat{H}^2 + \alpha'\hat{R}_{MNPQ}\hat{R}^{MNPQ} \right]$$
$$+ O(\alpha'^2) \tag{4}$$

One obvious solution to field equations corresponds to $\langle\hat{\phi}\rangle$ = constant, $\langle\hat{H}_{MNP}\rangle$ = 0 and $\langle\hat{g}_{MN}\rangle$ the flat metric but this is valid only for D = 26. In this case the possible ground states are given by

$$M_D = M^d \times T^k \tag{5}$$

when M^d is d-dimensional Minkowski space and T^k is the k-torus with d|= 26-k. However, for D > 26 the cosmological term in (4) obliges us to look for solutions in which some of the dimensions are compactified on a curved manifold and we can now follow the traditional Kaluza-Klein interpretation. Accordingly, we split in indices

$$x^M = (x^\mu, y^m) \tag{6}$$

where x^μ (μ = 1,...,d) refers to space-time and y^m (m = 1,...,k) to the extra dimensions. One solution which suggests itself corresponds to the case

$$M_D = M^d \times G \tag{7}$$

where G is a non-Abelian group manifold of dimension k given by d = D-k. In this case

$$\langle \hat{g}_{mn} \rangle = \frac{1}{c_A} f_{mpq} f_n{}^{pq} = g_{mn}(y) \tag{8}$$

where

$$f_{mpq} = L_m{}^i L_p{}^j L_q{}^k f_{ijk} \tag{9}$$

with f_{ijk} the structure constants and $L_m{}^i$ in the left-invariant Killing vectors. c_A is the second order Casimir in the adjoint representation. This will indeed be a solution to all orders in α' provided

$$\langle \hat{\phi} \rangle = \text{constant} \tag{10}$$

$$\langle \hat{H}_{mnp} \rangle = m f_{mnp} = H_{mnp}(y) \tag{11}$$

where m is a constant with the dimensions of mass which determines the size of the compact group manifold. How can we tell this yields a conformally invariant theory <u>to all orders in α'</u>? To see this we substitute the ground state values of $\hat{g}_{MN}$, $\hat{B}_{MN}$ and $\hat{\phi}$ into the string Lagrangian (3) to obtain

$$L_1 = \partial_+ x^\mu \partial_- x^\nu \eta_{\mu\nu} + \partial_+ y^m \partial_- y^n (g_{mn} + B_{mn}) \tag{12}$$

where

$$H_{mnp} = 3 \partial_{[m} B_{np]} \tag{13}$$

In (12) we have used the orthonormal gauge $\gamma_{ab} = e^\sigma \eta_{ab}$ and employed the co-ordinates $\xi_\pm = \sigma \pm \tau$. But (12) is nothing but the non-linear σ-model on $M^d \times G$ with Wess-Zumino term, a system well studied in the literature[5] and known to be conformally invariant provided we satisfy the critical dimension formula

$$D - 26 = \frac{k c_A}{2\rho + c_A} \tag{14}$$

and provided that the radius of the group manifold is quantized in units of $\alpha'^{-\frac{1}{2}}$:

$$\rho\, m^2 \alpha' = 1 \tag{15}$$

where p is an integer. The appearance of the integer p follows from the topological quantization condition on the coefficient of the Wess-Zumino term.

The case p = 1 is rather remarkable because, in this case, L_1 is entirely equivalent to

$$L_2 = \partial_+ x^\mu \partial_- x^\nu \eta_{\mu\nu} + i\,\lambda^i \partial_+ \lambda^i + i\,\psi^i \partial_- \psi^i \tag{16}$$

i.e., a system of free fermions where i runs over the vector representation of G. But L_2 is also entirely equivalent to

$$L_3 = \partial_+ x^\mu \partial_- x^\nu \eta_{\mu\nu} + \partial_+ y^I \partial_- y^I \tag{17}$$

where I = 1,...,r where r is the rank of G, i.e., a system of free bosons on the torus of dimension r. The only restriction on G is that it be simply laced. See Table 3. In this case

$$d = D - k = 26 - r.$$

In particular for $G = E_8 \times E_8$, $c_A = 60$, r = 16, D = 506 and d = 10. The Lagrangian L_3 simply corresponds to the $M^d \times T^r$ compactification, and this establishes the correspondence between the D = 506 Kaluza-Klein group-manifold approach of Duff, Nilsson and Pope[6] and the D = 26 Frenkel-Kac torus approach of Casher et al.[7] [In fact, the fermionization required in going to L_2 works only for the $SO(16) \times SO(16)$ subgroup of $E_8 \times E_8$ but the equivalence of L_1 and L_3 is unimpaired by this.]

Table 3

The simply-laced groups

G	dim G	r = rank G	c_A
SO(2r)	r(2r-1)	r	4r-4
SU(2r)	r(r+2)	r	2r+2
E_6	78	6	24
E_7	133	7	36
E_8	248	8	60

One advantage of the group manifold approach is that we can now immediately write down the Kaluza-Klein ansatz for the massless modes of the compactified theory. First, however, we should say a few words about Kaluza-Klein "consistency".

Since the isometry group of the group manifold is $G_L \times G_R$, and since the VEVs of $\hat{\phi}$ and $\hat{H}_{MNP}$ given above are also $G_L \times G_R$-invariant, the d-dimensional theory will contain the Yang-Mills gauge bosons of $G_L \times G_R$. For generic Kaluza-Klein theories, however, a consistent Kaluza-Klein ansatz for the massless sector requires that we keep only a subgroup of the full isometry group. A "consistent" ansatz is one for which all solutions of the d-dimensional theory are solutions of the original D-dimensional theory. For a generic Kaluza-Klein theory with homogeneous extra dimensions, a consistent ansatz retains all those fields and only those fields invariant under a transitively-acting subgroup K of the isometry group. In particular, for group manifolds the gauge bosons are only those of G_L and not those of the full isometry group $G_L \times G_R$. Moreover, the G_L ansatz is in general consistent only if we include Kaluza-Klein scalars in the symmetrized adjoint × adjoint representation of G_L. Experience with d = 11 supergravity, however, teaches us that there may exist certain exceptional theories where a consistent ansatz can be achieved without demanding this "K-invariance". This can happen either by including fields which are not K-invariant [e.g., the SO(8) ansatz for the S^7 compactification of d = 11 super-gravity] or by omitting fields which are [e.g., the omission of Kaluza-Klein scalars in the SO(3) ansatz on S^7]. See Refs. 7) and 14). In the present context, an example of the latter phenomenon was provided in Ref. 6), where we showed that the G_L ansatz for the bosonic string was consistent in spite of omitting Kaluza-Klein scalars. In Ref. 10) we went one step further and showed that even the full $G_L \times G_R$ ansatz is consistent, provided we pay the price of including scalars S^{ij} in the (adjoint G_L, adjoint G_R) representation. This somewhat confusing situation is summarized in Table 4. In presenting these results, we shall first give the ansatz for $G_L \times G_R$ without including the scalars and then indicate how their inclusion solves the problem of inconsistency.

<u>Table 4</u>

Gauge groups surviving in consistent truncations of theories compactified on the group manifold G, and the corresponding scalar representations.

	With KK scalars	Without KK scalars
Generic KK theory	G_L : $(adj_L \times adj_L)_{sym}$	—
Bosonic string theory	$G_L \times G_R$: (adj_L, adj_R)	G_L

Let us introduce the Killing vectors K^I on the group manifold G

$$K^I = \left(L^i, R^i \right) \tag{18}$$

where L^i are the generators of left translations

$$\left[L^i, L^j \right] = m f^{ij}{}_k L^k \tag{19}$$

and R^i are the generators of left translations

$$\left[R^i, R^j \right] = -m f^{ij}{}_k R^k \tag{20}$$

and

$$[L^i, R^j] = 0 \tag{21}$$

The corresponding Yang-Mills gauge potentials are denoted by

$$A^I = (A^i, \tilde{A}^i) \tag{22}$$

where A^i are the gauge bosons of G_L and $\tilde{A}_i$ the gauge bosons of G_R. The corresponding field strengths are given by

$$F^I = (F^i, \tilde{F}^i) \tag{23}$$

where

$$F^i = dA^i + \frac{1}{2} f^i{}_{jk} A^j \wedge A^k \tag{24}$$

$$\tilde{F}^i = d\tilde{A}^i - \frac{1}{2} f^i{}_{jk} \tilde{A}^j \wedge \tilde{A}^k \tag{25}$$

The Killing vector components L^i_a and R^i_a satisfy the Cartan-Maurer equations

$$\nabla_a L_b{}^i = -\frac{m}{2} f^i{}_{jk} L^j_a L^k_b \tag{26}$$

$$\nabla_a R_b{}^i = \frac{m}{2} f^i{}_{jk} R^j_a R^k_b \tag{27}$$

and we also introduce the notation

$$\begin{aligned}
f_{abc} &= f_{ijk} L^i_a L^j_b L^k_c \\
&= f_{ijk} R^i_a R^j_b R^k_c
\end{aligned} \tag{28}$$

We are now in a position to state the ansatz for $\hat{\phi}$, $\hat{g}_{MN}$ and $\hat{B}_{MN}$ and to calculate the corresponding curvatures and field strengths. For the scalar, we write

$$\hat{\phi}(x,y) = \text{constant} + \phi(x) \tag{29}$$

The metric ansatz is

$$\hat{g}_{\mu\nu} = g_{\mu\nu} + (A_\mu{}^i L^{m i} + \tilde{A}_\mu{}^i R^{m i})(A_\nu{}^j L^{n j} + \tilde{A}_\nu{}^j R^{n j})$$

$$\hat{g}_{\mu n} = (A_\mu{}^i L^{m i} + \tilde{A}_\mu{}^i R^{m i}) g_{m n}$$

(30)

$$\hat{g}_{m n} = g_{m n}$$

The antisymmetric tensor ansatz is

$$\hat{B}_{\mu\nu} = B_{\mu\nu} + (A_\mu{}^i L^{i m} \tilde{A}_\nu{}^j R^{j m} - A_\nu{}^i L^{i m} \tilde{A}_\mu{}^j R^{j m})$$

$$\hat{B}_{\mu n} = (A_\mu{}^i L^{m i} - B_\mu{}^i R^{m i}) g_{m n}$$

(31)

$$\hat{B}_{m n} = B_{m n}$$

where, in the absence of scalars, g_{mn} and B_{mn} are just the ground-state values given by (8) and (11).

The quantities $\phi(x)$, $g_{\mu\nu}(x)$ and $B_{\mu\nu}(x)$ will be interpreted as the scalar, metric and antisymmetric tensors in d-dimensional space-time, and the quantities $A_\mu^i(x)$ and $\tilde{A}_\mu^i(x)$ will be the Yang-Mills gauge bosons for G_L and G_R respectively. Equation (30) is just the "standard ansatz" for the metric tensor familiar to Kaluza-Klein theories, the novel feature is the ansatz for $\hat{B}_{MN}$, which also involves the Yang-Mills gauge bosons. In this Kaluza-Klein interpretation, the gauge symmetry $G_L \times G_R$ is just a subgroup of the d-dimensional general co-ordinate group. To see this in more detail, consider a general co-ordinate transformation

$$x^M \to x^M - \hat{\xi}^M$$

(32)

and the corresponding transformations of ϕ, g_{MN} and B_{MN}. Then focus one's attention on the very special transformation

$$\hat{\xi}^M(x,y) = \left(0 \; , \; \alpha^i(x) L^{i m}(y) + \tilde{\alpha}^i(x) R^{i m}(y) \right)$$

(33)

with α^i and $\tilde{\alpha}^i$ arbitrary. Then from the Kaluza-Klein ansätze (29)-(31), we may compute the transformation rules for the d-dimensional fields. We find not only the usual Yang-Mills transformation rules for ϕ, $g_{\mu\nu}$, A^i_μ and $\tilde{A}^i_\mu$, but also that the $B_{\mu\nu}$ field transforms as

$$\delta B_{\mu\nu} = A^i_\mu \, \partial_\nu \alpha^i - A^i_\nu \, \partial_\mu \alpha^i - \tilde{A}^i_\mu \, \partial_\nu \tilde{\alpha}^i + \tilde{A}^i_\nu \, \partial_\mu \tilde{\alpha}^i \tag{34}$$

It is now tedious but straightforward to compute the curvature $\hat{R}_{ABCD}$ and the field strength $\hat{H}_{ABC}$. In particular the d-dimensional field strength H is not just dB but rather

$$H = dB - m^{-2} \left(\Omega - \tilde{\Omega} \right) \tag{35}$$

where

$$\Omega = F^i \wedge A^i - \frac{1}{6} f^{ijk} A^i \wedge A^j \wedge A^k \tag{36}$$

$$\tilde{\Omega} = \tilde{F}^i \wedge \tilde{A}^i + \frac{1}{6} f^{ijk} \tilde{A}^i \wedge \tilde{A}^j \wedge \tilde{A}^k \tag{37}$$

i.e., we have acquired d-dimensional Chern-Simons terms, even though there were no such terms in D dimensions where $\hat{H} = d\hat{B}$. The non-covariance of Ω and $\tilde{\Omega}$ under Yang-Mills transformations is exactly cancelled by the unusual transformation rule for $B_{\mu\nu}$ of (34), ensuring that H does not transform. Hence, although in D dimensions $d\hat{H} = 0$, in d-dimensions we have (on using $\alpha' m^2 = 1$)

$$dH + \alpha' \left(F^i \wedge F^i - \tilde{F}^i \wedge \tilde{F}^i \right) = 0 \tag{38}$$

Note that left- and right-handed gauge fields enter with opposite sign and that this equation is exact to all orders in α'. This is the result quoted in the second line of Table 2.

In a similar fashion, we may substitute the Kaluza-Klein ansatz into the D-dimensional equations of motion and hence derive the corresponding d-dimensional equations. Here, however, we must be careful to ensure that the ansatz is consistent and, as explained in Ref. 10) this requires the inclusion of scalars into the ansatz which we have so far omitted for simplicity. Otherwise we obtain unacceptable constraints on the other massless fields like $F_{\mu\nu}^{i}\tilde{F}^{\mu\nu j} = 0$ arising from putting $S^{ij} = 0$ in the scalar field equation $\square S^{ij} \sim F_{\mu\nu}^{i}\tilde{F}^{\mu\nu j}$.

3. THE HETEROTIC AND TYPE II STRING

To obtain the corresponding terms for the heterotic string, we

* Choose $G = E_8 \times E_8$ for which dim $G = 496$, $r = 16$, $c_A = 60$, $D = 506$ and hence $d = 10$
* Decompose $G_R \supset SO(8)$, i.e., $\tilde{A}_{\mu}^{i}$ ($i=1,\ldots,496$) → $\tilde{A}_{\mu}^{ab}$ ($a,b=1,\ldots,8$);
* Following Ref. 9), identify the right-handed Yang-Mills gauge potential with the gravitational spin connection

$$\tilde{A}_{\mu}^{ab} = \omega_{\mu(+)}^{ab} \tag{39}$$

From (38), we obtain

$$dH = \alpha' \, Tr \left(R_+ \wedge R_+ - F \wedge F \right) \tag{40}$$

But these are just the heterotic string Yang-Mills and Lorentz Chern-Simons terms quoted in third line of Table 2.

Note that in (39) and (40), it is the spin connection with torsion $\omega_{(+)}$ which appears, where

$$\omega_{(+)} = \omega + \frac{1}{2} H \tag{41}$$

To understand this, consider the heterotic string σ-model[11] in the orthonormal gauge

$$L_H = \partial_+ x^\mu \partial_- x^\nu (g_{\mu\nu} + B_{\mu\nu}) + i\lambda^a (\partial_+ \lambda^a + \omega_{(+)\mu}^{ab} \lambda^b \partial_+ \xi^\mu)$$

$$+ i\psi^A (\partial_- \psi^A + A_\mu^{AB} \psi^B \partial_- x^\mu) + \frac{1}{2} F_{\mu\nu}^{AB} \lambda^\mu \lambda^\nu \psi^A \psi^B$$

$$(42)$$

where $\xi_\pm = \tau \pm \sigma$. If our previous claims are correct, we must be able to derive this from the bosonic string σ-model

$$L_B = \partial_+ x^M \partial_- x^N (\hat{g}_{MN} + \hat{B}_{MN}) \tag{43}$$

by (a) compactifying on the group manifold (b) substituting in the Kaluza-Klein ansatz (c) fermionizing the extra dimensional co-ordinates y^m (m = 1,...,k) and then making the identification (41). An interesting question is the origin of the four-fermi term in (42). This will be discussed elsewhere[15].

It should be admitted, however, that in common with Casher et al.[8] we have as yet no __dynamical__ understanding of the decomposition $G_R \supset SO(8)$ and the identification (39). Nor do we see any justification for the truncation of the string spectrum which seems to be entailed in reproducing that of the heterotic string. The idea is that states whose G_R index i (i = 1,...,496) runs over the 8_s __spinor__ index α (α = 1,...,8) of $SO(8)_{internal}$ transform as __fermion__ representations of the diagonal $SO(8)$

$$SO(8)_{heterotic} \subset SO(8)_{bosonic} \times SO(8)_{internal} \tag{44}$$

which is identified as the transverse space-time group of the heterotic string. Hence, in some sense, the G_R Yang-Mills boson $\tilde{A}_\mu^\alpha$ is really the gravitino

$$\widetilde{A}_\mu{}^\alpha \;\rightarrow\; \psi_\mu{}^\alpha \tag{45}$$

the scalars $S_i{}^\alpha$ are really the gauginos

$$S_i{}^\alpha \;\rightarrow\; \chi_i{}^\alpha \tag{46}$$

and the Yang-Mills parameters $\widetilde{\alpha}^\alpha$ is really the supersymmetric parameter

$$\widetilde{\alpha}^\alpha \;\rightarrow\; \varepsilon^\alpha \tag{47}$$

but the origin of the Fermi statistics remains obscure. We would like to be able to say that there are distinct vacua of the bosonic string relative to one of which all states transform as bosons, but relative to the other some states transform as fermions. Unfortunately, we are not yet in a position to make this more precise. Nevertheless, we glimpse the beginnings of the explanation for supersymmetry by using (39), (45) and (47) to convert the G_R Yang-Mills transformation rule

$$\delta \widetilde{A}_\mu{}^i \;=\; \partial_\mu \widetilde{\alpha}^i + f^i{}_{jk}\, \widetilde{A}_\mu{}^j\, \widetilde{\alpha}^k \tag{48}$$

into the gravitino transformation rule

$$\delta \psi_\mu{}^\alpha \;=\; \partial_\mu \varepsilon^\alpha - \frac{1}{4}\, \omega_\mu{}^{ab} \left[\gamma_{ab}\right]^\alpha{}_\beta\, \varepsilon^\beta + \dots \tag{49}$$

on using the property of the E_8 structure constants

$$f^\alpha{}_{[ab]\,\beta} \;=\; -\frac{1}{4} \left[\gamma_{ab}\right]^\alpha{}_\beta \tag{50}$$

Similarly the Type IIA and Type IIB theories are obtained by

* decomposing both $G_R \supset SO(8)$ and $G_L \supset SO(8)$
* identifying both the left- and right-handed Yang-Mills gauge potentials with gravitational spin connection

$$A_\mu{}^{ab} = \omega_{(-)\mu}{}^{ab} \quad , \quad \widetilde{A}_\mu{}^{ab} = \omega_{(+)\mu}{}^{ab} \tag{51}$$

to obtain the type II non-linear σ-model, with the Chern-Simons term

$$dH = \alpha' \, Tr \left(R_+ \wedge R_+ \; - \; R_- \wedge R_- \right) \tag{52}$$

which is just the final line of Table 2.

The extra 64 bosonic degrees of freedom are then provided by $S^{\dot\alpha\beta}$ or $S^{\alpha\beta}$ corresponding to the embeddings $(8_c, 8_s)$ in the case of Type IIA or $(8_s, 8)$ in the case of Type IIB. The second supersymmetry of Type II has the same origin in G_L as did the first in G_R, i.e., with $A_\mu{}^\alpha$ and α^α playing the part of gravitino and supersymmetry parameter.

Thus we arrive at the bizarre picture of a three-in-one world that can be described equivalently in 10, 26 or 506 dimensions as in Table IV.

TABLE 4:   A three-in-one world described equivalently by 10, 26, or 506 dimensions.

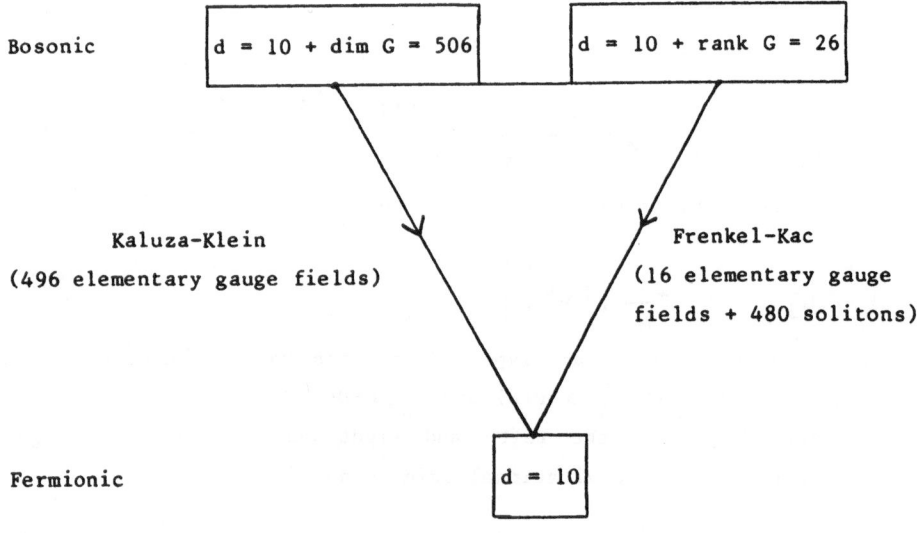

Bosonic

d = 10 + dim G = 506   d = 10 + rank G = 26

Kaluza-Klein   Frenkel-Kac
(496 elementary gauge fields)   (16 elementary gauge
                                      fields + 480 solitons)

Fermionic   d = 10

Of course, if the bosonic string really is the fundamental theory perhaps we should consider compactifications not from d to 10 dimensions but from d to four dimensions, i.e., on a group manifold for which from Eq. (1)

rank G = 22

But which G should we choose and why should the string prfer rank 22 to some other rank <26?

4. FOUR DIMENSIONAL SPACE-TIME FROM THE K3 LATTICE

The major unresolved problem of string theories is that of vacuum degeneracy. For although the higher-dimensional string equations have almost no parameters, all predictive power seems to be lost by the apparent multitude of different compactifications. In particular, it remains a mystery why the dimension of the uncompactified space-time should be just four. This section, based on a paper by Nilsson and myself[14], is an attempt to resolve some of these questions.

We begin by recalling the recent work of Narain[17] who considers compactifying into tori (10-d) and (26-d) dimensions of the right moving superstring and left moving bosonic string sectors respectively. Since the k-torus is given by R^k factored by a discrete lattice $\Gamma (T^k = R^k/\Gamma)$, the question devolves upon which Γ to choose. Narain points out that the condition for modular invariance is equivalent to self-duality for even Lorentzian lattices with (10-d) timelike and (26-d) spacelike directions. Let us denote such even self-dual lattices $\Gamma_{(p,q)}$, where (p,q) is the signature. Then all such Lorentzian lattices are iso-morphic to the lattice

$$n E_8 \oplus q P_2 \tag{53}$$

where

$$p - q = 8n \tag{54}$$

for some integer n, where E_8 is the root lattice of E_8 with Euclidean signature (8,0)

$$E_8 = \begin{bmatrix} 2 & -1 & 0 & 0 & 0 & 0 & 0 & 0 \\ -1 & 2 & -1 & 0 & 0 & 0 & 0 & 0 \\ 0 & -1 & 2 & -1 & 0 & 0 & 0 & 0 \\ 0 & 0 & -1 & 2 & -1 & 0 & 0 & 0 \\ 0 & 0 & 0 & -1 & 2 & -1 & 0 & -1 \\ 0 & 0 & 0 & 0 & -1 & 2 & -1 & 0 \\ 0 & 0 & 0 & 0 & 0 & -1 & 2 & 0 \\ 0 & 0 & 0 & 0 & -1 & 0 & 0 & 2 \end{bmatrix} \tag{55}$$

and where P_2 is a two-dimensional lattice with signature (1,1)

$$\begin{bmatrix} 0 & 1 \\ 1 & 0 \end{bmatrix} \tag{56}$$

corresponding to the group SU(2)×SU(2). This means that any even self-dual lattice $\Gamma_{(q+8n,q)}$ with q > 0 can be obtained from $nE_8 \oplus qP_2$ by means of an SO(8n+q,q) transformation. Distinct compactifications are then characterized by points in the coset

$$SO(p,q) / SO(p) \times SO(q)$$

There is a recent theorem due to Freedman[18] that states that all even self-dual Lorentzian lattices are given by the "intersection form" of a simply-connected topological four-manifold M^4. See the article by Stern[17] for a readable introduction to this branch of mathematics. Such a manifold will have Euler number

$$\chi = 2 + b_2 \tag{57}$$

where the second Betti number b_2 counts the number of harmonic two-forms. If we denote these two-forms by α_i (i = 1,...,b_2) then the intersection form is defined by

$$\Gamma_{ij} = \int_{M_4} \alpha_i \wedge \alpha_j \tag{58}$$

and obviously has rank b_2. Its signature (p,q) is given by (b_2^+,b_2^-), and the Hirzebruch signature by

$$\tau = b_2^+ - b_2^- \tag{59}$$

where b_2^+ count the number of self-dual two-forms and b_2^- the number of antiselfdual. (Hence τ must be a multiple of 8.) So the question of which is the right vacuum has been replaced by which is the right four-manifold.

Now Freedman's theorem involves <u>topological</u> four-manifolds but not every four-manifold is <u>differentiable</u>. For example, a necessary condition due to Rochlin[20] is that τ must be a multiple of 16. Suppose, just for fun, we use the criterion of <u>differentiability</u> of the four-manifold to narrow down the choice of vacuum.

Unfortunately, the question of which four-manifolds are smoothable (i.e., differentiable) is an outstanding problem in mathematics, but some very interesting results are known. For example

$$E_8 \oplus E_8$$

is not, even though $\tau = 16$! (This result, due to Donaldson[21], has created quite a stir in mathematical circles because it lies at the heart of the proof that R^4 has more than one differentiable structure.) P_2 on the other hand corresponds to the intersection form on $S^2 \times S^2$ which is differentiable. The problem then is to determine whether

$$2E_8 \oplus q\, P_2$$

is smoothable for some $q > 0$.

Note that from the string point of view, our criterion of differentiability then means that we cannot go from $D = 26$ to $d = 10$ but may perhaps go to $d < 10$! The amazing fact is that the case $q = 3$ is differentiable and simply corresponds to the four-manifold K3 for which $b_2^+ = 19$, $b_2^- = 3$ and $b_2 = 22$.

K3 is defined as a quartic surface in complex projective three-space CP^3 by

$$K3 = \left\{ [z_0, z_1, z_2, z_3] \in \mathbb{C}P^3 : z_0^4 + z_1^4 + z_2^4 + z_3^4 = 0 \right\}$$

Applying Narain's techniques using this very special lattice leads to a heterotic string theory with unique space-time dimension $d = 7$. The low-energy limit corresponds to $d = 7$, $N = 2$ supergravity coupled to super-Yang-Mills with rank 19 gauge group $E_8 \times E_8 \times SU(2) \times SU(2) \times SU(2)$. [The remaining rank 3 gauge group simply corresponds to the three U(1)'s of

N = 2 supergravity.] Corresponding theories in d = 10-q < 7 could also be obtained by taking the topological sum of K3 and (q-3) copies of $S^2 \times S^2$. Thus a four-dimensional theory could be obtained from

$$\Gamma_{(22,6)} = \Gamma[K3] \oplus 3\,\rho_2$$

whose low-energy limit is N = 4 supergravity coupled to the rank 22 gauge group $E_8 \times E_8 \times [SU(2)]^6$. [Once again, the remaining rank six group simply corresponds to the six U(1)'s of N = 4 supergravity.] Unfortunately, the "minimal" theory is in d = 7 and there seems no compelling reason for adding three $S^2 \times S^2$ manifolds to K3.

So far we have followed Narain and considered only the heterotic string, but the situation becomes much more interesting if we adopt the point of view that the fundamental theory is the _bosonic_ string. Once again we must compactify on a torus factored by an even self-dual Lorentzian lattice but now with signature (26-d, 26-d). The "minimal" theory in the sense described above is now given by

$$\Gamma_{(22,22)} = \Gamma[K3] \oplus \Gamma[\overline{K3}]$$

where $\overline{K3}$ corresponds to the four-manifold obtained from K3 by reversing the orientation and has $b_2^+ = 3$, $b_2^- = 10$ and $\tau = -16$. Hence

$$d = 26 - 22 = 4 \tag{60}$$

and we obtain a _four-dimensional_ bosonic string with gauge group G×G, where G is the rank 22 group $E_8 \times E_8 \times SU(2)^6$.

Thus our objective is now to repeat the derivation of superstrings from bosonic strings discussed in Section 1 but now compactifying from D = 26 to d = 4 on the torus T^{22} defined by the intersection form of the four-manifold K3+$\overline{K3}$ [or, bearing in mind the previously discussed equivalence, from D = 518 to four on the group manifold $E_8 \times E_8 \times SU(2)^6$]. However, the outcome is no longer clear. In particular, it is unclear whether a chiral N = 1 theory would result. If a chiral theory does not emerge directly in this way, it may be necessary to go one stage further and compactify not merely on the torus T^{22} defined by K3+$\overline{K3}$ but on $T^{22}/\mathcal{G}$ where $\mathcal{G}$ is a discrete group. Factorings of T^6 by $\mathcal{G}$ have been considered by Dixon et al.[22] but to obtain chiral fermions, it was necessary that $\mathcal{O}_{\mathcal{G}}$ had fixed points thus leading to singularities, i.e., "orbifolds". From our K3 point of view, a more attractive possibility is that advocated by Lam and Li[23] who consider direct compactification from 26 to 4 via $T^{22}/\mathcal{O}_{\mathcal{G}}$ where $\mathcal{O}_{\mathcal{G}}$ acts on T^{22} without fixed points, so that $T^{22}/\mathcal{O}_{\mathcal{G}}$ is a genuine manifold without singularities. These authors claim to obtain chiral N = 1 theories in this way while still preserving modular invariance. (They consider $E_8 \times E_8 \times SU(3)^3$ rather than $E_8 \times E_8 \times SU(2)^6$.]

The vital question remaining is why string theory should select intersection forms of _differentiable_ manifolds, but if it does it would explain why we cannot remain in d = 10. And in answer to the question why space-time has four dimensions we would reply: because the second Betti number of K3 equals 22!

ACKNOWLEDGEMENTS

I am grateful for conversations with A. Chamseddine, B. Nilsson, C. Pope, D. Ross and N. Warner.

REFERENCES

1) Green, M.B. and Schwarz, J.H., Phys. Lett. B149, 117 (1984).

2) Freund, P.G.O., Phys. Lett. B151, 387 (1985).

3) Frenkel, I. and Kac, V.G., Inv. Math. 62, 23 (1980);
   Goddard, P. and Olive, D., in "Workshop on Vertex Operators in Mathematics and Physics", Berkeley (1983).

4) Gross, D., Harvey, J., Martinec, E. and Rohm, R., Phys. Rev. Lett. 54, 502 (1985); Nucl. Phys. B256, 253 (1985).

5) Witten, E., Comm. Math. Phys. 92, 455 (1984);
   Nemeschensky, D. and Yankielowicz, S., Phys. Rev. Lett. 54, 620 (1984);
   Altschüler, D. and Nilles, H.P., Phys. Lett. 154B, 135 (1985);
   Goddard, P. and Olive, D., Nucl. Phys. B257, 226 (1985);
   Jain, S., Shankar, R. and Wadia, S.R., Phys. Rev. D32, 2713 (1985);
   Bergshoeff, E. Randjbar-Daemi, S., Salam, A., Sarmadi, H. and Sezgin, E., Nucl. Phys. B269, 77 (1986).

6) Duff, M.J., Nilsson, B.E.W. and Pope, C.N., Phys. Lett. B163, 343 (1985), also published in Proc. Cambridge Workshop on Supersymmetry and its applications (June-July 1985), (Eds. Gibbons, Hawking and Townsend, C.U.P. 1986).

7) Duff, M.J., Nilsson, B.E.W. and Pope, C.N., Physics Reports $\underline{130}$, 1 (1986).

8) Casher, A., Englert F., Nicolai, H. and Taormina, A., Phys. Lett. $\underline{B162}$, 121 (1985); see also Englert, F., Nicolai, H. and Schellekens, A., CERN preprint TH.4360/86 (1986).

9) Charap, J.M. and Duff, M.J., Phys. Lett. $\underline{B69}$, 445 (1977).

10) Duff, M.J., Nilsson, B.E.W., Pope, C.N. and Warner, N.P., Phys. Lett. $\underline{171B}$, 170.

11) Hull, C.M., Nucl. Phys. $\underline{B267}$, 266 (1986).

12) Duff, M.J., in Proceedings of the GR11 Conference, Stockholm, July 1986, CERN preprint TH.4568/86.

13) Duff, M.J., in Proceedings of the 1985 Les Houches Summer School (Eds. Ramond and Stora).

14) de Wit, B. and Nicolai, H., CERN preprint TH.4359/86 (1986).

15) Chamseddine, A., Duff, M.J., Nilsson, B.E.W., Ross, D. and Pope, C.N., in preparation.

16) Duff, M.J. and Nilsson, B.E.W., Phys. Lett. $\underline{175B}$, 417 (1986).

17) Narain, K.S., Phys. Lett. $\underline{B169}$, 41 (1986).

18) Freedman, M., Diff. J. Geom. $\underline{17}$, 357 (1983).

19) Stern, R.J., The Mathematical Intelligencer $\underline{5}$, 39 (1985).

20) Rochlin, V.A., Dokl. Akad. Nauk SSR $\underline{84}$,221 (1952).

21) Donaldson, S.K., Bull. Amer. Math. Soc. $\underline{8}$, 81 (1983).

22) Dixon, L., Harvey, J.A., Vafa, C. and Witten, E., Nucl. Phys. $\underline{B261}$, 678 (1985).

23) Lam, C.S. and Da-Xi Li, McGill University preprints (1985).

NON LINEAR EFFECTS IN QUANTUM GRAVITY

Ian Moss
Department of Theoretical Physics
University of Newcastle upon Tyne
Newcastle upon Tyne   NE1 7RU
U.K.

ABSTRACT

Canonical quantum gravity can be reduced in a semi-classical limit to conventional quantum gravity on a curved spacetime background. Changes in the topology of space require a reformulation of the theory which introduces density matrices or non-linear terms into the semi-classical limit.

1. INTRODUCTION

We are still in the prehistory of a quantum theory of gravity. I shall report here how recent investigations into the origin of the universe, stimulated by the success of the inflationary scenario [1] as an explanation of the large scale structure of the universe, has lead to the development of new ideas in quantum cosmology. In particular, we shall see how the Schrodinger equation is recovered from quantum gravity and how changes in the topology of spacetime can fundamentally influence quantum theory and its interpretation.

In constructing a quantum model of the universe we need to introduce a fundamental action and initial conditions. The gravitational part of the action presents particular difficuties. We shall use the Einstein-Hilbert action for the time being. It may be that the theory based upon this action can be rescued from some apparent inconsistences. In any case, we expect that we have a good approximation whenever the radius of curvature of spacetime is larger than the Planck length of 10^{-33} cm. This is analogous to the use of the Coulomb potential in describing a Hydrogen atom where we fix a boundary condition on the wave function at the centre, despite the fact that we know that the Coulomb potential is invalid inside of the nucleus.

For initial conditions we shall make use of Hawking's suggestion that "spacetime is finite but unbounded" [2] . This is realised by the Hartle-Hawking prescription [3] for the quantum state of the universe. This state is a function of the geom-

geometry of 3-dimensional hypersurfaces Σ described by a metric tensor g_{ij} and matter fields ϕ. The state is defined by

$$\psi(g,\phi) \;=\; \sum e^{\,i\,S[^{4}g,\,^{4}\phi]} \tag{1}$$

where we sum over all 4-geometries and matter configurations such that the 4-geometry is compact and has no boundary other than Σ (fig. 1).

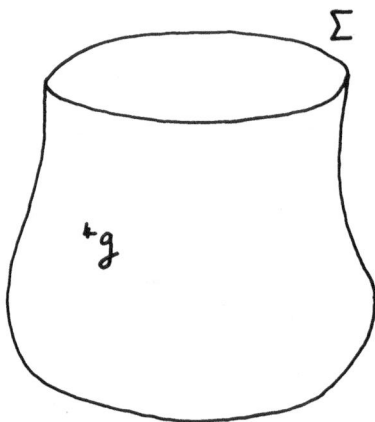

Figure 1

Approximate calculations [4,5,6] of this wave function in various inflationary models has demonstrated that it is a superposition of states representing universes with a satisfactory large scale structure. This means that they are spatially flat and homogeneous with scale-free density fluctuations. We shall discuss such a decomposition of the wave function in sect.2 . We are confronted, however, with considerable problems of interpretation. The observer is necessarily part of the system as in the "Many Worlds" interpretation of quantum mechanics [7]. In this picture, the collapse of the wave function associated with a measurement becomes a splitting of the wave function into non-interacting branches. With quantum cosmology this leaves us with a problem : which universe from the superposition do we live in and what causes the splitting? Furthermore, the wave function gives us probabalistic information, but the meaning of probability is unclear when we have just one unrepeatable experiment.

A remarkable relationship between changes in spatial topology and these questions will be explained in sect. 3.

2. CANONICAL QUANTUM GRAVITY

In the canonical approach to quantum gravity we decompose spacetime into 3-dimensional spatial hypersurfaces Σ_t . The phase space consists of 3-metrics g_{ij} and various matter fields ϕ on Σ_t, together with their conjugate momenta p^{ij} and π. Under this canonical decomposition the action has the form

$$S = \int (\dot{g} p + \dot{p} \pi - NH - N_i H^i) \, d^4x \qquad (2)$$

where the indices on g and p are implicit. The superhamiltonian H can be decomposed into gravitational and matter parts,

$$H (g, p, \varphi, \pi) = H_g (g, p) + H_m (g, \varphi, \pi) \qquad (3)$$

Einstein's theory of gravity corresponds to the choice

$$H_g (g, p) = G_{ijkl} p^{ij} p^{kl} - \sqrt{g} R(g) \qquad (4)$$

$$H_g^i (g, p) = -2 p^{ij}{}_{|j} \qquad (5)$$

where bar denotes the g-covariant derivative with Ricci scalar R, and G_{ijkl} is the DeWitt metric [8],

$$G_{ijkl} = \frac{1}{2} \frac{1}{\sqrt{g}} (g_{ij} g_{kl} + g_{il} g_{jk} - g_{ij} g_{kl}) \qquad (6)$$

The lapse and shift functions N and N_i act as Lagrange multipliers for the constraints H=0 and H^i=0, which can be viewed as a subset of Einstein's equations. The other Eintein equations follow from the variation of the action (2) with respect to g and p. They can be expressed in terms of covariant derivatives on configuration space with a metric $\mathcal{G} = (NG_{ijkl}, NG_{\varphi\varphi})$. Classical solutions are represented by trajectories in configuration space which are geodesics when the matter fields are massless.

In the quantum theory states can be represented by wave functions

$\Psi(g, \phi)$. The constraints must be realised by [8,9]

$$H \psi = 0 \qquad (7)$$

$$H^i \psi = 0 \qquad (8)$$

with p replaced by $i\delta/\delta g$. Eq. 7 is known as the Wheeler-DeWitt equation. There is a non-trivial factor ordering problem associated with this equation. We shall choose a factor ordering which uses the covariant derivative [10] . This gives

$$H_g = -\nabla_g^2 + U \tag{9}$$

where $U = g^{\frac{1}{2}} R(g)$ and

$$\nabla_g^2 = G^{-\frac{1}{2}} \frac{\delta}{\delta g_{ij}} G_{ijkl} G^{\frac{1}{2}} \frac{\delta}{\delta g_{kl}} \tag{10}$$

This factor ordering implies that the Wheeler-DeWitt equation is invariant under coordinate redefinitions on the configuration space.

If we choose instead to quantise the theory by path integrals, then the transition amplitude is given by

$$\langle g', \varphi' | g, \varphi \rangle = \int d[g, \varphi] e^{iS[^4 g, \varphi]} \tag{11}$$

where the 4-geometry ^4g interpolates between the 3-metrics g and g'. This amplitude satisfies the Wheeler-DeWitt equation provided that we use the configuration space metric G to define the path integral measure,

$$d[g, \varphi] = \prod_x (\det G)^{\frac{1}{2}} dg_{ij}(x) d\varphi(x) dN(x) dN_i(x) \tag{12}$$

which is invariant under coordinate redefinitions.

Because of the vanishing of the superhamiltonian H it is impossible to introduce a $\partial\psi/\partial t$ term on the right hand side of the Wheeler-DeWitt equation, unlike the normal Schrodinger equation. Instead, the Wheeler-DeWitt equation is a dynamical equation because H forms a hyperbolic operator. This is only possible because H is not a positive definite Hamiltonian.

The fact that ψ does not depend upon time is simply an expression of general covariance, because time is a coordinate label. Physical questions about time development have to be addressed by choosing some degrees of freedom to form clock subsystems against which the time development of the remaining system can be measured.

When the gravitational field behaves semi-classically, then it is possible to measure the passing of time by the evolving geometry. In this limit it has been shown in special cases, by DeWitt [8] and Banks [11] that the Wheeler-DeWitt equation reduces to the familiar Schrodinger equation. We shall generalise their results to include the back reaction of the matter fields on the geometry.

Consider a wave function of the form

$$\psi(g,\varphi) = \psi_g(g)\,\psi_m(\varphi,\tau) \tag{13}$$

The parameter τ, defined below, becomes the time coordinate. We shall construct a gravitational wave function ψ_g which is sharply peaked around a solution of Einstein's equations with a back reaction term and then demonstrate that ψ_m satisfies Schrodinger's equation.

The semi-classical gravitational field must satisfy a Hamilton-Jacobi equation,

$$H_g\left(g,\frac{\delta\bar{S}}{\delta g}\right) + \langle H_m \rangle = 0 \tag{14}$$

where $\langle H_m \rangle$ represents the back reaction of matter, given by

$$\langle H_m \rangle = \int d[\varphi]\,\psi_m^{\dagger}(\varphi,\tau)\,H_m(g,\varphi,i\delta/\delta\varphi)\,\psi_m(\varphi,\tau) \tag{15}$$

The time coordinate is defined by

$$\frac{\delta g_{ij}}{\delta\tau} = G_{ijkl}\frac{\delta\bar{S}}{\delta g_{kl}} \tag{16}$$

where G_{ijkl} is the DeWitt metric. As may be expected, the principal function $\bar{S}$ can be identified with the gravitational action of the background field, including the back reaction term $\langle H_m \rangle$.

The semi-classical approximation is given by

$$\psi_g(g) = A(g)\,e^{i\bar{S}(g)} \tag{17}$$

where A is a slowly varying function of g. When substituted into the Wheeler-DeWitt equation we neglect the $\partial^2 A/\partial g^2$ terms, but no others. The leading order terms vanish due to eq. 14, and the next order terms vanish for the choice

$$A = \left(\det\frac{\delta^2\bar{S}}{\delta g\,\delta g_0}\right)^{\frac{1}{2}}\exp\left(i\int\langle H_m\rangle\,d^3x\,d\tau\right) \tag{18}$$

where g_0 are integration constants. The remaining terms give Schrodinger's equation,

$$H_m \psi_m = i \delta \psi_m / \delta \tau \qquad (19)$$

This equation has to be solved in the fixed semi-classical background metric g, which in turn depends upon the wave function ψ_m.

A general solution of the Wheeler-DeWitt equation will develop into a superposition of WKB components, with coefficients which give an idea of the relative importance of each semi-classical metric g. The difficulty in describing how an observer can reduce the wave function to one of these WKB components has alwready been mentioned in the introduction.

3. CHANGES IN TOPOLOGY

We could imagine transition amplitudes between disconected metrics, such as the one shown in fig. 2 which represents $\langle g_1, g_2 | g \rangle$. In particle theory, the analagous diagram would be an interaction vertex which indicates the necessity for second quantisation. In the case of the amplitude $\langle g_1, g_2 | g \rangle$, the Wheeler-DeWitt equation breaks down because the transition between the initial and final geometries is discontinuous. We can use the path integral instead, but there are theorems in differential topology which imply that no Lorentzian 4-geometry exists which can match on to topologically distinct 3-geometries. This is a fundamental difference between Euclidean and Lorentzian formulations of the path integral.

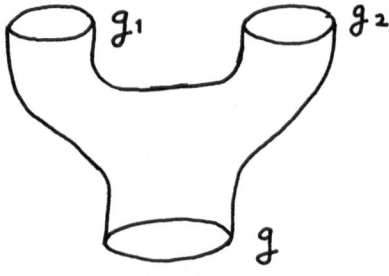

Figure 2

There are two distinct ways in which we can proceed:

(i) Sum over all of the unobservable components of the
    3-geometry

(ii)Extend the Hilbert space to include $|g\rangle$ + $|g_1 g_2\rangle$ + ..

Case (i) is based upon an idea of Hawking [12] and Page [13] , and
leads to transitions from pure states to density matrices. Consider
the joint transition amplitude P($|g_1\rangle \rightarrow |g_1'\rangle$, $\langle g_2| \rightarrow \langle g_2'|$). Allowing
for changes in topology with a sum over the unobserved components,

$$ P = \langle g_1'|g_1\rangle\langle g_2|g_2'\rangle + \sum_g \langle g_1',g|g_1\rangle\langle g_2|g,g_2'\rangle + \ldots \tag{20} $$

We can represent this diagramatically by fig. 3, where g is the metric
on an internal surface shown by the dotted line. Each diagram corresponds to a Euclidean path integral with boundary metrics g_1, g_2, g_1', g_2'.

A pure state has the form ρ = $|\psi\rangle\langle\psi|$ where $|\psi\rangle$ = $\sum_g \psi(g) |g\rangle$. The
development of this state is described by a superscattering operator S,
where

$$ \$\rho = \sum_{g_1 g_2} \sum_{g_1' g_2'} \psi(g_1)\psi(g_2)|g_1\rangle\langle g_2| P(|g_1\rangle\langle g_2| \rightarrow |g_1'\rangle\langle g_2'|) \tag{21} $$

From the expansion of P given in eq. 20 we can see that this expression
does not factorise in general and therefore $\$\rho$ does not represent a
pure state.

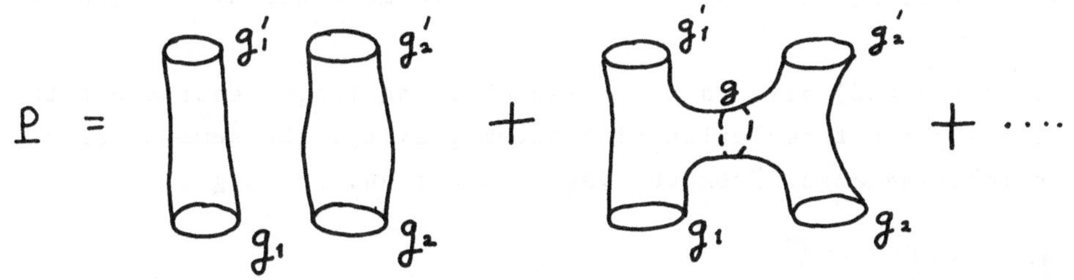

$$ P = \underbrace{}_{} \;\; + \;\; \underbrace{}_{} \;\; + \;\; \cdots $$

Figure 3

In quantum cosmology, the development of pure to mixed states can
be interpreted as a branching of the wave function. For example, a pure
state $|RW1\rangle$ + $|RW2\rangle$ representing a superposition of Robertson-Walker
universes 1 and 2 can evolve to the density matrix $|RW1\rangle\langle RW1|$ + $|RW2\rangle\langle RW2|$

in which all of the quantum interference terms between the universes vanish. This may free us from having to involve a complicated observer in order to split the wave function.

Case (ii) involves a second quantisation of the Wheeler-DeWitt equation. We proceed by constructing an action whose classical equations of motion are the constraint equations. Next, we introduce interaction terms corresponding to fig. 2 and finally, the theory is quantised by path integration over a suitable class of wave functions ψ.

Some of the gauge freedom can be fixed at this stage, introducing ghost fields θ and θ^i into the wave function $\psi(g,\theta,\theta^i)$. The remaining BRS symmetry imposes a constraint $Q\psi = 0$, which replaces the constraints (7) and (8). Using the methods of Fradkin and Vilkoviski [14] we can construct Q in terms of H, H^i and their commutators, with the result that

$$Q = \int d^3x \left\{ H\theta + H_i \theta^i + \hat{\theta} \nabla_i \theta \theta^i + \tfrac{1}{2} \hat{\theta}_i (\nabla_k \theta^i) \theta^k - \tfrac{1}{2} \hat{\theta}_i (\nabla_k \theta^k) \theta^i \right\} \tag{22}$$

The anticommuting fields $\hat{\theta}$ and $\hat{\theta}_i$ are antighost fields which are canonically conjugate to θ and θ^i.

A suitable action is given by

$$S = \int d[g,\theta,\theta^i] \left(\psi Q \psi + \tfrac{2}{3} \psi \psi \psi \right) \tag{23}$$

Variation of the first term in the action leads to the "free" constraint equation $Q\psi = 0$ which is equivalent to the Hamiltonian and momentum constraints. The second term contains cubic interaction terms as shown in fig. 2, but quartic and higher terms could also be included.

The action (23) gives a theory which is no longer equivalent to the Wheeler-DeWitt formulation of quantum gravity. The semi-classical limit is modified also. From the variation of eq. 23 we get

$$Q\psi + \psi\psi = 0 \tag{24}$$

From a decomposition of the wave function analagous to eq. 13, we no longer obtain Schrodinger's equation but we get a non linear generalisation of it. The particular form of this equation depends upon how the product of ψ fields is defined. For the simplest choice ($\psi(g,\theta))^3$, the result is that

$$H_m \psi_m = i\, \delta\psi_m/\delta\tau - \psi_3\, \psi_m^2 \qquad (25)$$

It is well known that such non-linear terms can induce effects which resemble the collapse of the wave function [15] . This can happen when there exist solitonic solutions of eq. 25 which typical solutions to the linearised equation can evolve into. In the case of quantum cosmology, it may be possible for the selection of a universe to occur without the intervention of an observer.

Non-linear terms can also arise in an analagous situation where a many electron wave function is approximated by the wave function of a single electron. An example of this would be the Hartree-Fock approximation for a Helium atom, in which a single electron is viewed as if it where moving in its own charge distribution. Such non-linear effects are also important in solid state physics, where the soliton solutions representing photons have been observed [16] .

4. REFERENCES

1.   A.H. Guth, Phys. Rev. D23 (1981) 347.

2.   S.W. Hawking, Pontif. Accad. Sci. Varia, 48 (1982) 563.

3.   J.B. Hartle and S.W. Hawking, Phys. Rev. D23 (1983) 2960.

4.   I.G. Moss and W.A. Wright, Phys. Rev. D29 (1984) 1067.

5.   S.W. Hawking and J. Luttrell, Nuc. Phys. B247 (1984) 250.

6.   I.G. Moss, "The New Cosmogony", to appear in the proceedings of the IV Marcel Grossman Meeting, Rome 1985.

7.   B.S. DeWitt and N. Graham, eds. "The Many Worlds Interpretation of Quantum Mechanics", Princeton University Press 1973.

8.   B.S. DeWitt, Phys. Rev. 160 (1967) 1113.

9.   J.A. Wheeler, in "Battelle Rencontres", eds. C. DeWitt and J.A. Wheeler, Benjamin New York, 1968.

10. S.W. Hawking and D.N. Page, Nuc. Phys. B264 (1986) 185.

11. T. Banks, Nuc. Phys. B249 (1985) 332.

12. S.W. Hawking, "The density matrix of the universe" (Cambridge preprint 1986).

REFERENCES (CONT.)

13. D.N. Page, "Density matrix of the universe"
    (Pennsylvania preprint 1986).

14. E.S. Fradkin and G.A. Vilkoviski, Phys. Lett. 55B (1975) 224.

15. D.Bohm and J. Bub, Rev. Mod. Phys. 38 (1966) 453.

16. A.R. Bishop and T. Schneider, eds. "Solitons and Condensed
    Matter Physics" (Springer-Verlag, Berlin 1978).

OUR UNIVERSE AS AN ATTRACTOR IN A SUPERSTRING MODEL

Kei-ichi MAEDA

International Centre for Theoretical Physics, Trieste, Italy[*]

Abstract: One preferential scenario of the evolution of the universe is discussed
in a superstring model. The universe can reach the present state as an attractor
in the dynamical system. The kinetic terms of the 'axions' play an important
role so that our present universe is realized almost uniquely.

I. INTRODUCTION

A superstring theory is a promising candidate for a fundamental unified
theory including gravity[1]. It may be successful from the phenomenological point
of view[2]. Its application to cosmology is certainly important and interesting.
The superstring theory as well as the other unified theories such as the Kaluza-Klein
idea[3] predict a higher-dimensional space-time, which may play a very important role
in the early universe. Our world is, however, definitely four dimensional at least
in the macroscopic scale. The 4-dimensional Hot Big Bang scenario is very
successful. We beleive the Friedmann expanding universe based on the 4-dimensional
Einstein gravity. Hence, if we take a higher-dimensional space-time seriously, we
must explain how our 4-dimensional universe is naturally realized in the higher-
dimensional space-time.

The present universe must be [the 4-dim Friedmann universe (F^4)] x [a very
small static internal space (K)][4]. In the conventional 4-dim theory, the isotropy
and the homogeneity of space-time, which may be deduced from the cosmological principle
or from an inflationary scenario[5], guarantee that our universe is a Friedmann space-
time. In a higher-dimensional theory, however, that is not true because the F^4 x K
space-time is not isotropic at all in higher dimensions. Our anisotropic universe

----

* Address after October 1986 : Observatoire de Paris-Meudon, Groupe d'Astrophysique
  Relativiste, 92195 Meudon, France

(F^4 x K) must be a special space-time in the dynamical system.   Namely, the F^4 x K

solution should be <u>an attractor</u> in our system.   If this attractor is strong enough

to guarantee for the universe to reach the F^4 x K solution for a wide range of initial

conditions, we understand easily why our universe is now in the present state[6].

For example, in the 6-dim, N=2 supergravity model, the F^4 x K space-time is a unique

attractor and all the space-time (apart from the time reversal ones) approach F^4 x K

asymptotically in the later stage of the universe[7].

The second problem in a higher-dimensional theory is that the reduced four-

dimentional effective gravity theory may not be the Einstein theory but the Jordan-

Brans-Dicke (JBD) theory for a Ricci-flat compactification such as a Calabi-Yau

manifold.   The JBD parameter is given by the dimension D of the internal space as

$\omega = -(D-1)/D$, then this theory should be excluded from the astrophysical observations

($\omega > 500$).

Thirdly, since inflation is very desirable in modern cosmology[5], we should

also search our unified theory for inflation.   In general, this task is not so easy

in unified theories, because we cannot add an inflaton responsible for inflation by

hand as we like.   We have been already given a set of fields and we must look for

the inflaton among them.

Hence, the present main problems in higher-dimensional unified theories, from

the cosmological point of view, are:

   (i)   Can the 4-dim Friedmann universe be realized naturally as an attractor in the

         higher-dimensional space-time ?

   (ii)   Can the 4-dim Einstein gravity be obtained from a higher-dimensional theory,

         rather than the JBD theory ?

 (iii)   Does inflation really occur in the unified theory ?

We investigate the above problems and discuss one preferential scenario in the

10-dim, N=1 supergravity model with E_8 x E_8' Yang-Mills fields and the additional

curvature squared terms, both of which are derived from the heterotic string model

in the field theory limit[8].   The effective 4-dim Lagrangian is given in §.II ,

assuming a Ricci-flat compactification.   In §.III, we consider one simple model

without fermion condensations and show that the F^4 x K is always a unique attractor

in this system, but the effective gravity theory is the JBD theory with $\omega = -1$.   In

§.IV, we take into account a gluino condensation of E_8' gauge fields, which is

responsible for the local SUSY breaking.   We show that the minimum of the potential, which corresponds to the $F^4 \times K$ space-time because of zero cosmological constant, is always one of the attractors if the 3-space is expanding. One preferential scenario in our model and remarks on inflation are discussed in §.V.

II. FOUR-DIMENSIONAL LAGRANGIAN

Assuming a Ricci-flat (e.g. a Calabi-Yau) compactification, the 10-dimensional world interval is described by[F1]

$$ds_{10}^2 = e^{\Phi/2} \, \bar{g}_{\mu\nu} \, dX^\mu dX^\nu = e^{\Phi/2} [\, b^{-6} ds_4^2 + b^2 \, d\tilde{s}_6^2 \,] \tag{2.1}$$

with

$$ds_4^2 = g_{mn}(x) \, dx^m dx^n \quad \text{and} \quad d\tilde{s}_6^2 = \tilde{g}_{MN}(y) \, dy^M dy^N. \tag{2.2}$$

The conformal factor $\exp(\Phi/2)$ is from the Weyl rescaling of the 10-dim metric in the string action[8]. $\Phi(x)$ is the dilaton and $b(x)$ is the 'radius' of the internal space. We have factorized out the conformal factor b^{-6} in the 4-dim metric in order to obtain the proper Einstein action in four dimensions. $\tilde{g}_{MN}$ is the metric of a static Ricci-flat manifold.

The bosonic part of the 10-dim, N=1 supergravity Lagrangian[9] consists of

$$\mathscr{L}_B = \mathscr{L}_R + \mathscr{L}_\Phi + \mathscr{L}_{F^2} + \mathscr{L}_{H^2} \tag{2.3-a}$$

with

$$\mathscr{L}_R = \frac{\sqrt{-\bar{g}}}{2\kappa_{10}^2} R(\bar{g}) \quad , \quad \mathscr{L}_\Phi = -\frac{\sqrt{-\bar{g}}}{4\kappa_{10}^2} (\bar{\nabla}\Phi)^2 ,$$

$$\mathscr{L}_{F^2} = -\frac{\sqrt{-\bar{g}}}{4 g_{10}^2} e^{-\Phi/2} \, \text{Tr} \, F_{\mu\nu}^2 \quad \text{and} \quad \mathscr{L}_{H^2} = -\frac{3\kappa_{10}^2}{4 g_{10}^4} \sqrt{-\bar{g}} \, e^{-\Phi} H_{\mu\nu\rho}^2 , \tag{2.3-b}$$

where $\kappa_{10}^2 / 8\pi = G_{10}$ and g_{10} are the 10-dim gravitational constant and gauge coupling constant, respectively, whilst $R(\bar{g})$ and $\bar{\nabla}$ are the scalar curvature and the covariant derivative with respect to $\bar{g}_{\mu\nu}$.

In the case of a Calabi-Yau compactification, we need the Riemann curvature squared term[2], which is derived in the field-theory limit of a superstring theory[8]. Here, we assume the special combination of curvature squared terms;

$$R_{\mu\nu\rho\sigma}^2(\bar{g}) - 4 R_{\mu\nu}^2(\bar{g}) + R^2(\bar{g}) \tag{2.4}$$

in order to have a ghost-free theory[10]. Through the vacuum expectation value (VEV) of Yang-Mills fields;

$$F_{\mu\nu} = \begin{cases} \widetilde{F}_{MN}(y) & \text{for } \mu,\nu = M,N \\ 0 & \text{otherwise} \end{cases} \tag{2.5}$$

we obtain a Calabi-Yau compactification in the non-static background (2.1)[11-13].
The curvature squared terms and $\mathcal{L}_{F^2}$ are rewritten as

$$\mathcal{L}_{R^2} + \mathcal{L}_{F^2} = -\frac{\sqrt{-\widetilde{g}}}{4\widetilde{g}_{10}^2} e^{-\Phi/2} \left\{ b^{-4}(Tr \, \widetilde{F}_{MN}^2 - 30 \, \widetilde{R}_{MNPQ}^2) \right\}$$
$$+ I(g_{mn}, b, \Phi) + \text{(totally divergent term)}, \tag{2.6}$$

where $\widetilde{R}_{MNPQ}$ is the Riemann tensor with respect to $\widetilde{g}_{MN}$. The first term in Eq.(2.6)
vanishes because of the Calabi-Yau compactification. $I(g_{mn}, b, \Phi)$ is the term from
$\mathcal{L}_{R^2}$ which depends only on g_{mn}, b and Φ, and it does not contain higher-order
time derivatives of g_{mn}, of b and of Φ. If the time scale or length scale of
changes in g_{mn}, b and Φ is much smaller than the Planck scale (e.g. in the later
stage of the universe), then the I-term can be neglected as compared with the other
terms such as $\mathcal{L}_R$[11]. It is worth noting that this may not be always true for
some combinations of curvature squared terms. Because the structure of the dynamical
system may change completely if higher-order time derivatives appear[11,13]. The
expression (2.6) is also valid for the simple torus compactification with vanishing
$\widetilde{F}_{MN}$ and $\widetilde{R}_{MNPQ}$.

The Einstein action is reduced in four dimensions as

$$\int d^{10}X \, \mathcal{L}_R = \frac{1}{2\kappa^2} \int d^4x \, \sqrt{-g} \left[R(g) - 24 (\nabla \ln b)^2 \right], \tag{2.7}$$

where $\kappa^2 = \kappa_{10}^2 / \int d^6y \sqrt{\widetilde{g}}$, and R(g) and ∇ are the scalar curvature and the covariant
derivative with respect to g_{mn}.

The VEVs of H_{mnp} and of its potential B_{MN} provide two 'axions', θ_S and θ_T
defined by

$$H_{mnp} = \frac{1}{3\kappa} e^{-2\sqrt{2}\kappa\phi_S} \epsilon_{mnpq} \nabla^q \theta_S, \text{ and } B_{MN} = \frac{1}{\sqrt{3}\kappa} \theta_T \epsilon_{MN}. \tag{2.8}$$

Here, we introduce the new scalar fields ϕ_S and ϕ_T, instead of Φ and $\ln b$, as

$$\phi_S = \frac{1}{\sqrt{2}\kappa}(6 \ln b - \Phi/2) \text{ and } \phi_T = \frac{\sqrt{6}}{2\kappa}(2 \ln b + \Phi/2). \tag{2.9}$$

(ϕ_S, θ_S) and (ϕ_T, θ_T) form two complex chiral superfields S and T in four dimensions,
defined by

$$S = e^{\sqrt{2}\varkappa\phi_S} + i\sqrt{2}\varkappa\theta_S \quad \text{and} \quad T = e^{\sqrt{\frac{2}{3}}\varkappa\phi_T} + i\varkappa\theta_T/\sqrt{3} \; . \tag{2.10}$$

Using ϕ_S and ϕ_T , the 10-dim world interval (2.1) is written as

$$dS_{10}^2 = e^{-\sqrt{2}\varkappa\phi_S} dS_4^2 + e^{\sqrt{\frac{2}{3}}\varkappa\phi_T} d\tilde{S}_6^2 \; . \tag{2.11}$$

The VEVs of the internal components of H_{MNP} may also appear through Dirac string singularities for a non-simply connected interal manifold such as a Calabi-Yau manifold. As for the VEVs of fermions, we consider only the gluino χ condensation of E_8' gauge field, which may give natural SUSY breaking mechanism[14]. This mechanism with the above VEVs of H_{MNP} provides the effective 4-dimensional potential:

$$V(\phi_S, \phi_T, \theta_S) = \frac{4\pi^2}{\varkappa^4} e^{-\varkappa(\sqrt{2}\phi_S + \sqrt{6}\phi_T)}$$
$$\times \left| c - h\, e^{-\frac{3\sqrt{2}}{2b_0}i\varkappa\theta_S} \exp\left(-\frac{3}{2b_0} e^{\sqrt{2}\varkappa\phi_S}\right) \right|^2 \; , \tag{2.12}$$

where contants c and h are defined by

$$H_{IJK} = c\, m_{PL}^3\, \epsilon_{IJK} \quad \text{and} \quad Tr\, \bar{\chi}\, \Gamma_{IJK}\chi \sim h\mu^3\epsilon_{IJK} \; , \tag{2.13}$$

respectively. m_{PL} and μ are the Planck mass and the energy scale of condensation. b_0 is fixed by the gauge group.

From the above setting, we obtain the four dimensional effective Lagrangian,[15] which is equivalent to that of the 10-dim, N=1 supergravity model, as

$$S = \int d^4x \sqrt{-g}\, L \tag{2.14-a}$$

$$L = \frac{1}{2\varkappa^2} R(g) - \frac{1}{2}[(\nabla\phi_S)^2 + (\nabla\phi_T)^2 + e^{-2\sqrt{2}\varkappa\phi_S}(\nabla\theta_S)^2 + e^{-2\sqrt{\frac{2}{3}}\varkappa\phi_T}(\nabla\theta_T)^2]$$
$$- V(\phi_S, \phi_T, \theta_S) \; . \tag{2.14-b}$$

III. THE FRIEDMANN UNIVERSE AS AN ATTRACTOR

We, now, consider the cosmological solutions. The 4-dim metric ds_4^2 is assumed to be

$$dS_4^2 = g_{mn}\, dx^m\, dx^n = -dt^2 + a^2(t)\, dx^2 \; . \tag{3.1}$$

The basic equations are

$$3H^2 = \kappa^2 \left[E_{kin} + V + \rho \right] , \tag{3.2}$$

$$2\dot{H} = -\kappa^2 \left[2E_{kin} + P + \rho \right] , \tag{3.3}$$

$$\ddot{\phi}_S + 3H\dot{\phi}_S + \sqrt{2}\,\kappa\, e^{-2\sqrt{2}\kappa\phi_S}\dot{\theta}_S^2 + \frac{\partial V}{\partial \phi_S} = 0 , \tag{3.4}$$

$$\left(a^3 e^{-2\sqrt{2}\kappa\phi_S}\dot{\theta}_S \right)^{\cdot} + a^3 \frac{\partial V}{\partial \theta_S} = 0 , \tag{3.5}$$

$$\ddot{\phi}_T + 3H\dot{\phi}_T + \sqrt{\tfrac{2}{3}}\,\kappa\, e^{-2\sqrt{\frac{2}{3}}\kappa\phi_T}\dot{\theta}_T^2 + \frac{\partial V}{\partial \phi_T} = 0 , \tag{3.6}$$

and

$$\left(a^3 e^{-2\sqrt{\frac{2}{3}}\kappa\phi_T}\dot{\theta}_T \right)^{\cdot} = 0 , \tag{3.7}$$

where $H = \dot{a}/a$ is the Hubble parameter, $E_{kin} = (\dot{\phi}_S^2 + \dot{\phi}_T^2 + e^{-2\sqrt{2}\kappa\phi_S}\dot{\theta}_S^2 + e^{-2\sqrt{\frac{2}{3}}\kappa\phi_T}\dot{\theta}_T^2)/2$, and a dot denotes the derivative with respect to t. Here, we have also introduced the 4-dim matter fluid, which energy density and pressure are denoted by ρ and P, respectively.

In this section, we consider the case without fermion condensations. we also assume that $H_{MNP} = 0$ (i.e. c = 0). The potential V vanishes. This might be justified if H_{MNP} is induced only when the gluinos condense as discussed by Rohm and Witten[16].

If V = 0, we can obtain the analytic solutions of Eqs.(3.2~7) as follows. Introducing the new time coordinate η instead of t by

$$d\eta = a^{-3} dt , \tag{3.8}$$

Eqs.(3. 4 - 3.7) are easily integrated as

$$\frac{d\theta_S}{d\eta} = Q_S\, e^{2\sqrt{2}\kappa\phi_S} , \qquad \frac{d\theta_T}{d\eta} = Q_T\, e^{2\sqrt{\frac{2}{3}}\kappa\phi_T} \tag{3.9}$$

$$\frac{1}{2}\left(\frac{d\phi_S}{d\eta}\right)^2 + \frac{1}{2} Q_S^2\, e^{2\sqrt{2}\kappa\phi_S} = E_S , \tag{3.10-a}$$

and

$$\frac{1}{2}\left(\frac{d\phi_T}{d\eta}\right)^2 + \frac{1}{2} Q_T^2\, e^{2\sqrt{\frac{2}{3}}\kappa\phi_T} = E_T , \tag{3.10-b}$$

where Q_S, Q_T, E_S and E_T are integration constants. Integrating these equations again, we obtain the analytic solutions as

$$e^{2\sqrt{2}\times\phi_S} = \frac{2E_S}{Q_S^2} \Big/ ch^2\left[2x\sqrt{E_S}\,(\eta-\eta_o')\right] \tag{3.11-a}$$

$$e^{2\sqrt{\frac{2}{3}}\times\phi_T} = \frac{2E_T}{Q_T^2} \Big/ ch^2\left[2x\sqrt{\frac{E_T}{3}}\,(\eta-\eta_o'')\right] \tag{3.11-b}$$

$$\theta_S = \theta_{S,o} + \frac{\sqrt{E_S}}{x\,Q_S}\,\tanh\left[2x\sqrt{E_S}\,(\eta-\eta_o')\right] \tag{3.12-a}$$

and
$$\theta_T = \theta_{T,o} + \frac{\sqrt{3E_T}}{x\,Q_T}\,\tanh\left[2x\sqrt{\frac{E_T}{3}}\,(\eta-\eta_o'')\right] \tag{3.12-b}$$

for $Q_S \neq 0$ and $Q_T \neq 0$. For $Q_S = Q_T = 0$,

$$\phi_S = \phi_{S,o} \pm \sqrt{2E_S}\,\eta\,, \tag{3.13-a}$$

$$\phi_T = \phi_{T,o} \pm \sqrt{2E_T}\,\eta\,, \tag{3.13-b}$$

$$\theta_S = \theta_{S,o} \quad\text{and}\quad \theta_T = \theta_{T,o} \tag{3.14}$$

Here, η_o', η_o'', $\theta_{S,o}$, $\theta_{T,o}$, $\phi_{S,o}$, and $\phi_{T,o}$ are integration constants.

The Einstein equations are, now,

$$3H_\eta^2 = x^2\left[E + P a^6\right]\,, \tag{3.15}$$

and
$$2\frac{dH_\eta}{d\eta} = -x^2(P-\rho)\,a^6\,, \tag{3.16}$$

where $H_\eta = (da/d\eta)/a$ and $E = E_S + E_T$. Using the above solutions (3.11-3.14), we finally obtain a complete set of solutions for the following two cases:

CASE (I) : Vacuum ($P = \rho = 0$)

$$a = a_o \exp\left[\pm x\sqrt{\frac{E}{3}}\,\eta\right]$$
$$t = \pm\frac{a_o^3}{x\sqrt{3E}}\,e^{\pm x\sqrt{3E}\,\eta} \tag{3.17}$$

CASE (II) : $P = (\gamma - 1)\rho$ ($0 < \gamma < 2$)

$$a = \left[\frac{x^2 E}{\delta}\Big/ sh^2\left\{\frac{(2-\gamma)x}{2}\sqrt{3E}\,(\eta-\eta_o)\right\}\right]^{\frac{1}{3(2-\gamma)}}$$
$$t \sim \pm\left(\frac{2-\gamma}{\gamma}\right)\left[\frac{4}{3\delta(2-\gamma)^2}\right]^{\frac{1}{2-\gamma}}|\eta-\eta_o|^{\frac{\gamma}{2-\gamma}} \quad\text{for}\quad \eta\to\eta_o\neq 0 \tag{3.18}$$

,

where a_0 and η_0 are integration constants and the constant δ is defined by

$$\delta = \chi^2 \rho \, a^{3\delta}.$$ (3.19)

For the case (I), $\eta \to \infty$ when $t \to \infty$. The scale factor a expands as $a \propto t^{1/3}$, which is the same as that of the stiff matter ($P = \rho$) dominated universe, because of the massless scalar fields (ϕ_S, θ_S, ϕ_T and θ_T). θ_S and $\theta_T \to$ const, but ϕ_S and $\phi_T \to -\infty$ ($\pm\infty$ for $Q_S = Q_T = 0$). The internal space shrinks to zero, hence this asymptotic solution is not $F^4 \times K$.

For the case (II), $t \to \infty$ when $\eta \to \eta_0 - 0$. $a \to t^{2/3\delta}$ asymptotically, which is the Friedmann universe with matter fluid of $P = (\delta - 1)\rho$. Since $\phi_S, \phi_T, \theta_S, \theta_T \to$ some constants, the internal radius approaches some constant. All solutions approach the $F^4 \times K$ space-time for $t \to \infty$. Therefore, the Friedmann universe with a static internal space is a unique attractor in our dynamical system.

From the above solutions, we find one important role of Q_S and Q_T (the kinetic terms of the 'axions')[17]. Eq.(3.10) shows that there are maximum values for ϕ_S and ϕ_T if $Q_S \neq 0$ and $Q_T \neq 0$, i.e.

$$\phi_S \leq \frac{1}{2\sqrt{2}\chi} \ln\left(\frac{2E_S}{Q_S^2}\right) \quad \text{and} \quad \phi_T \leq \frac{\sqrt{3}}{2\sqrt{2}\chi} \ln\left(\frac{2E_T}{Q_T^2}\right).$$ (3.20)

Those kinetic terms provide potential barriers which prevent ϕ_S and ϕ_T from going away to arbitrary large values. This result is important, because the universe must stay near the preferential minimum ($\sim$ the Planck scale) when the gluinos condense, to reach the present state for natural initial conditions.

The effective 4-dim action (2.14) contains the proper Einstein action, then it seems that this model guarantees the Einstein gravity. That is, however, not true because the original coupling of a massless scalar to the 4-dim gravity was removed from the action by the Weyl rescaling (the factor b^{-6}), but appeared in the 4-dim world interval (2.11). We need a potential for ϕ_S, which is the massless JBD scalar in our model, so that it fixes the value of ϕ_S and guarantees the 4-dim Einstein gravity.

IV. THE EINSTEIN GRAVITY AS AN ATTRACTOR

When the universe expands and the temperature drops below some critical value ($\sim \mu$), the gluinos of the largest gauge group (e.g. E_8') condense. We find the gluino-condensation potential[14]. As discussed by Rohm and Witten[16], the VEVs of H_{MNP} may be also induced through an instanton solution at the gluino condensation and its value c is quantized as

$$C = C_n \equiv n\,C_0 \; , \tag{4.1}$$

where n is an integer and c_0 is determined by the geometry of the internal space. If we take into account the Chern-Simon term, however, we find an instanton solution, through which the quantized value c_n can change to the other value (c_{n+1}). Bubbles may be formed through the quantum tunnelling. Since c changes through the quantum tunnelling, the potential V also changes with time. The time scale of quantum tunnelling ($\sim t_{QT}$) is not yet known because the explicit instanton solution has not been obtained. In this section, we shall discuss the evolution of the universe mainly for the case that $t_{QT} \gg 1/H$ (CASE (I)) and briefly for the other two cases $\left[t_{QT} \ll 1/H \; (\text{CASE (II) }) \text{ and } t_{QT} \sim 1/H \; (\text{CASE (III) }) \right]$. (See Ref.(12) for the details).

CASE (I): $t_{QT} \gg 1/H$

In this case, we can neglect the quantum tunnelling effect. c_n is actually constant, then the potential V is fixed during the evolution of the universe. The potential minima are located at

$$\phi_S = \phi_{S,0} \equiv \frac{1}{\sqrt{2}\,\chi}\, \ln\left(\frac{2 b_0}{3}\, \ln\left| \frac{h}{c} \right| \right) \tag{4.2-a}$$

$$\theta_S = \theta_{S,m} \equiv \frac{2\sqrt{2}\, b_0\, \pi}{3\,\chi}\, m \quad \text{(m: any integer)} \tag{4.2-b}$$

$$\phi_T = \text{arbitrary} \tag{4.2-c}$$

Since the potential V vanishes at these minima, it guarantees the $F^4 \times K$ as well as the 4-dim Einstein gravity. The analytic solutions are given by Eqs.(3.11-b,3.12-b, and 3.17 or 3.18) with $\phi_S = \phi_{S,0}$ and $\theta_S = \theta_{S,m}$. Since these solutions are a part of the previous solutions, all solutions approach the $F^4 \times K$ space-time as discussed before. The value of ϕ_S is fixed at $\phi_{S,0}$, then the theory at the low energy scale is effectively the Einstein gravity theory.

The potential V, however, has another unpreferential minimum at $\phi_S = \infty$, as shown in Fig. 1 for the case of $\theta_S = \theta_{S,m}$. Neither the Friedmann solution nor the Einstein gravity is not obtained at this minimum. Then, the following question arises. Which minimum is obtained for natural initial conditions of the universe ?

In order to answer this question, we must investigate the dynamics of the universe for general initial conditions.

Here, we shall show that the preferential minima $(\phi_{s,0}, \theta_{s,m})$ are always attractors in our dynamical system if the 3-space is expanding, and those can be reached with finite probability.

Let us introduce the new time coordinate by

$$d\tau = \exp\left(-\sqrt{\tfrac{3}{2}} \kappa \phi_T\right) dt \qquad (4.3)$$

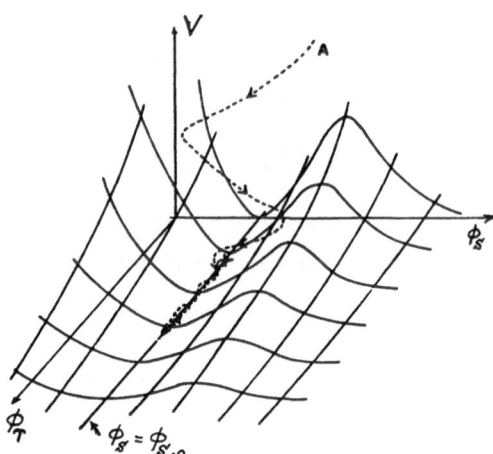

Fig. 1

Define the 'energy' ρ_s and the 'potential' $U(\phi_s, \theta_s)$ of the dynamical system for ϕ_s and θ_s by

$$\rho_s = \tfrac{1}{2} \left[\left(\frac{d\phi_s}{d\tau}\right)^2 + e^{-2\sqrt{2}\kappa\phi_s}\left(\frac{d\theta_s}{d\tau}\right)^2 \right] + U(\phi_s, \theta_s) \qquad (4.4\text{-a})$$

$$U(\phi_s, \theta_s) = V(\phi_s, \phi_T = 0, \theta_s) . \qquad (4.4\text{-b})$$

From Eqs.(3.4 and 3.5), the equation for ρ_s is written as

$$\frac{d\rho_s}{d\tau} = -3\zeta \left[\left(\frac{d\phi_s}{d\tau}\right)^2 + e^{-2\sqrt{2}\kappa\phi_s}\left(\frac{d\theta_s}{d\tau}\right)^2 \right] , \qquad (4.5\text{-a})$$

$$\zeta \equiv H_\tau - \kappa\left(\frac{d\phi_T}{d\tau}\right)/\sqrt{6} , \qquad (4.5\text{-b})$$

with $H_\tau = (da/d\tau)/a$. The constraint equation (3.2) reads

$$|H_\tau| > \kappa \left|\frac{d\phi_T}{d\tau}\right| /\sqrt{6} . \qquad (4.6)$$

If the 3-space is expanding (i.e. $H_\tau > 0$ or $H_\tau(\tau_0) > 0$ at some epoch $\tau = \tau_0$), ζ is always positive, then the system is always dissipative ($d\rho_s/d\tau < 0$). The minima $(\phi_{s,0}, \theta_{s,m})$ of the potential U are isolated. (The schematic shape is shown in Fig.2 for $\theta_s = \theta_{s,m}$). Therefore, once the universe is trapped in the shaded region T_E in Fig.2 at any value of ϕ_T, the universe always approaches the preferential minimum along the dotted lines A in Figs. 1 and 2. On the other hand, if the universe reaches the region T_{JBD}, then the universe always goes away to ϕ_s = infinity, finding the JBD theory, as the dotted line B in Fig.2 .

It is worth noting that the 'energy' of the total system, ρ_T, which is defined by

$$\rho_T = E_{kin} + V \qquad (4.7)$$

is also decreasing with time if the 3-space is expanding, i.e.

$$\dot{\rho}_T = - 6 H E_{kin} < 0 \qquad (4.8)$$

for $H > 0$. By losing the 'energy' ρ_T the universe reaches either the region T_E or the region T_{JBD} at some value of ρ_T. The minima ($\phi_{s,o}$, $\theta_{s,m}$) are always attractors in our system.

In Fig.3, we show the phase diagram of (ϕ_s, $\dot{\phi}_s$) for the case of $\theta_s = \theta_{s,m}(Q_s=0)$. The present universe ($\phi_{s,o}$, 0) is a nice attractor. If $\phi_s \lesssim$ (a few) $\times$ $\phi_{s,o}$ at the gluino condensation, then the universe may reach the preferential minimum [18]. If we take into account the kinetic terms of the 'axions' (Q_s and Q_T), the ϕ_s-field could stay near $\phi_s = \phi_{s,o}$ when the gluinos

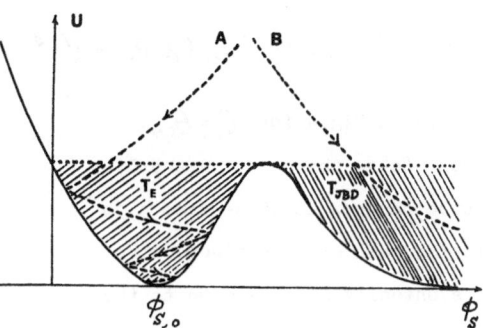

Fig. 2

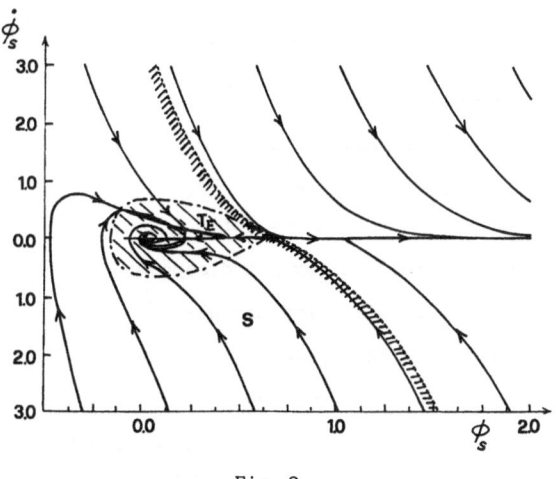

Fig. 3

condense for natural initial conditions, as discussed in §.III. Hence, the universe can reach the present state (the Friedmann universe and the 4-dim Einstein gravity) for a wide range of initial conditions.

We shall give brief comments for the cases (II) and (III).

CASE (II): $t_{QT} \ll 1/H$

The phase transition may occur due to the quantum tunnelling immediately after the universe will find the lower potential. For example (see Fig.4), the universe, which starts with the n=1 vacuum, reaches the point A, beyond which the potential with n=2 becomes lower than that with n=1. Then, the phase transition occurs just after passing through the point A. Many small bubbles with the n=2 vacuum are formed in the old n=1 vacuum, and collide with each other. The phase changes to n=2. The effective potential for the ϕ_s - and θ_s-fields, under which the universe evolves, is the minimum of all possible potentials, U_{eff}, i.e.

$$U_{eff}(\phi_s, \theta_s) \equiv min \left\{ U_n(\phi_s, \theta_s) \equiv V(\phi_s, \phi_T = 0, \theta_s; c = c_n) \,\Big|\, n \in Z \right\}. \qquad (4.9)$$

U_{eff} is shown in Fig.4 for $\theta_s = \theta_{s,m}$.
At each cusp (e.g. the points A and B),
the universe changes it phase from n to
n+1.   In the case of this figure, the
universe eventually settles down to the
n=3 vacuum.   If the universe reaches
one of the minima of U_{eff}, the $F^4 \times K$
space-time is realized asymptotically
and the 4-dim Einstein gravity is also
guaranteed as discussed in the case (I).

CASE (III): $t_{QT} \sim 1/H$

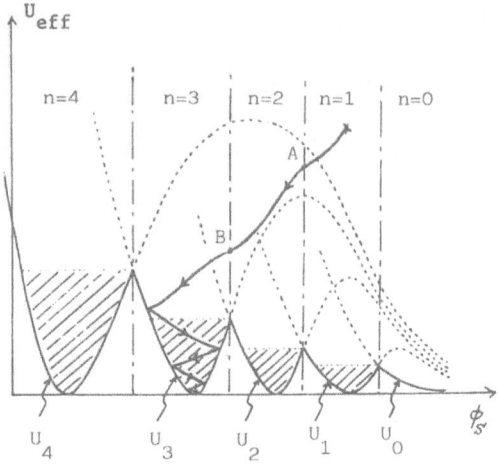

Fig. 4

In this case, we expect that
bubbles are formed at each cusp of U_{eff}
and the bubble structure may not disappear
before the universe will find the next
transition point.   For example, the (n_0+1)-
vacuum bubbles, which are formed in the initial
n_0 vacuum, evolve under the potential U_{n_0+1} and
reach the next transition point B.   The (n_0+2)-
vacuum bubbles are formed in the (n_0+1)-bubbles.
The old n_0 vacuum evolves under U_{n_0} and will
also find another transition point.   The
similar bubble formation occurs.   Finally, the
universe may find itself with a hierarchical
bubble structure as shown in Fig.5.   Each
bubble may reach some of the minima of U_{eff}.

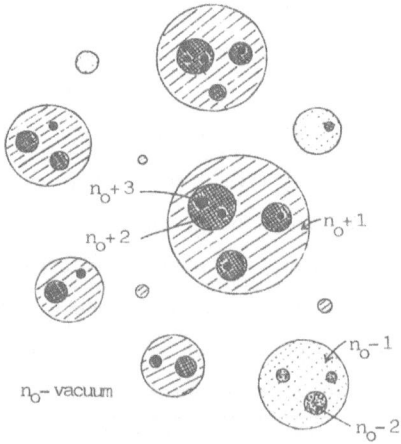

Fig. 5

The Newtonian gravitational constant G_N and the gauge coupling constant g_4 are
given by

$$G_N = e^{\sqrt{2}\kappa\phi_{s,0}} \kappa^2 \Big/ 8\pi \quad \text{and} \quad g_4^2 = e^{-\sqrt{2}\kappa\phi_{s,0}}, \qquad (4.10)$$

hence G_N and g_4 take different values in each bubble.   This bubble structure may
give rise to the so-called 'domain wall problem', unless inflation occurs after the

universe reaches one of the minima of the gluino-condensation potential V.

V. SUMMARY AND REMARKS

Since there is no kinetic term of the metric of the 10-dim 'target space' in a tree-level string action, we do not know yet the dynamics of the universe at the string level. We do not know yet what are the basic equations for the universe at the Planck time or beyond that stage. There are some indications of interesting phenomena[19], although many things are not clear yet.

Therefore, here we have discussed only the dynamics of the universe after the Planck time, in which period the field-theory limit may be justified. The dynamics of the universe is described by the Einstein equations with small corrections such as the curvature squared terms.

We summerize our scenario. If the compactification takes place near the Planck time and the gluinos condense below that scale, then we have two eras; the era before the gluino condensation and the era after that. In the first period, the space-time always approaches the $F^4 \times K$ solution, but the effective theory is the JBD gravity theory. If we take into account the kinetic terms of the 'axions', the scalar fields ϕ_S and ϕ_T may stay near the preferential minimum when the gluinos condense. After the gluino condensation, the universe may find itself in the present state (the Friedmann expanding universe and the 4-dim Einstein gravity) naturally for a wide range of initial conditions. Our universe may be obtained uniquely as <u>an attractor</u>.

The VEVs of H_{MNP} , however, may appear before the gluino condensation too, i.e. at the compactification. The appeared potential V has no minimum except for at infinity. Does this change the above scenario completely ? It seems that the scalar fields ϕ_S and ϕ_T go away to infinity , where V approaches zero, before the gluinos condense. The universe cannot stay near the preferential minimum $\phi_S = \phi_{S,o}$ and cannot reach it as discussed in §.IV. Our universe may not be realized for natural initial conditions.

However, if we take into account the effect of the 'axions', θ_S and θ_T, on the equations for the ϕ_S - and ϕ_T - fields, then we can easily show that the potential barrier appears and it prevents the universe from rolling down to ϕ_S , ϕ_T = infinity as follows[17]. When h = 0, Eq.(3.5) can be integrated as before, and then the equations for ϕ_S and ϕ_T are written as

$$\ddot{\phi}_S + 3H\dot{\phi}_S + \frac{\partial}{\partial \phi_S}(V + V_Q) = 0 \tag{5.1}$$

$$\ddot{\phi}_T + 3H\dot{\phi}_T + \frac{\partial}{\partial \phi_T}(V + V_Q) = 0 \quad , \tag{5.2}$$

where V_Q is defined by

$$V_Q = \frac{1}{2a^6}\left[\, Q_S^2\, e^{2\sqrt{2}\kappa\phi_S} + Q_T^2\, e^{2\sqrt{\frac{2}{3}}\kappa\phi_T} \,\right] \tag{5.3}$$

This effective potential V_Q provides a potential barrier against $\phi_S, \phi_T \to \infty$. The scalar fields ϕ_S and ϕ_T cannot go away to infinity because of this potential barrier. Therefore, the universe will stay in some finite region. If the kinetic energy of the 'axions' at the compactification is of order of unity in the Planck unit, the ϕ_S-field may stay near the Planck scale ($\sim \phi_{S,0}$). This potential barrier V_Q is propotional to a^{-6}, then it will disappear later when the universe expands enough.

When the gluinos condense, the universe is still staying around the minima $\phi_S = \phi_{S,0}$, then will reach one of the preferential minima. The probability for the universe to find itself in the present state may become very high. Our universe is realized almost uniquely as an attractor. Therefore, the VEVs of H_{MNP} before the gluino condensation do not change our scenario so much.

As mentioned in Introduction, inflation is one of the most desirable mechanism for the solution of the flatness, horizon and entropy problems in modern cosmology[5]. Can we find natural inflation in the present model ? The answer is, unfortunately, " NO ", so far. Here, we shall look at the difficulties.

We know, so far, two types of inflation; i.e. one is the potential type such as the GUT inflation[20] and the other is that due to the curvature squared terms[21]. Recently, inflation of Kaluza-Klein type is also proposed for higher-dimensional theories[22]. We search our model for such inflations.

Kaluza-Klein inflation due to a rapid contraction of the internal space does not work in our model. Because this rapid contraction is caused by the internal curvature, which vanishes in the Ricci-flat compactification. KK inflation due to the curvature squared terms also does not work. Because the flat potential for the internal radius is generated again from the internal curvature and the cosmological constant, both of which do not exist in the present model.

Inflation of Starobinski type is not possible. Because if the curvature squared (R^2-) terms are the so-called Gauss-Bonnet combination (2.4)[10], those do not provide the R^2-terms in 4 dimensions. Even if the R^2-terms turn out to be the other combination[23], inflation is impossible due to the coupling to the dilaton field, as discussed the details in Ref.(24). The similar situation happens for the 10-dim de Sitter solution[25], i.e. the coupling to the dilaton destroys the de Sitter type solutions. Forthermore, the R^2-terms from the string theory have ambiguity depending on the renormarization scheme.[23] Hence, the cosmological application of the R^2-terms (except for the Riemann curvature squared term) might be meaningless.

As for inflation of the potential type, we have not known yet what could be the inflaton. We have not found natural inflation. It might also be difficult because of the absense of free (or small) dimensionless parameter in the string theory, although there are a few proposals[26]. Whether natural inflation can be obtained in the string theory is one of the most important questions in the super-string cosmology.

ACKNOWLEDGEMENTS

The author would like to thank P.Y.T. Pang, M.D. Pollock and C.E. Vayonakis, with whom some part of the present work has been done. He is also grateful to Professor N. Dallaporta and Professor D.W. Sciama for their kind hospitality at the International School for Advanced Studies, Trieste and to Professor Abdus Salam, the International Atomic Energy Agency and UNESCO for hospitality at the International Centre for Theoretical Physics, Trieste.

F1 : The indecies $\mu, \nu, \ldots$ run from 0 to 3 and 5 to 10, while m,n,... run from 0 to 3. M,N,... are used for the internal indecies. Our signature and notations are the same as those of " GRAVITATION ", by C. Misner, K.S. Thorne and J.A. Wheeler (Freeman, San Francisco, 1973).

REFERENCES

(1) J.H. Schwarz, Phys. Report 89(1982) 233; preprint, CALT-68-1290(1985);
       M.B. Green, Surveys in High Energy Physics 3(1983) 127;
       D.J. Gross, J.A. Harvey, E. Martinec and R. Rohm, Phys. Rev. Lett. 54(1984)
          502; Nucl. Phys. B256(1985) 253; B267(1986) 75.

(2) P. Candelas, G.T. Horowitz, A. Strominger and E. Witten, Nucl. Phys. B258
          (1985) 46.

(3) Th. Kaluza, Sitz. Preuss Acad. Wiss. K1(1921) 966;
       O. Klein, Z. Phys. 37(1926) 895; Nature 118(1926) 516.

(4) W.J. Marciano, Phys. Rev. Lett. 52(1984) 489.

(5) A.D. Linde, Rep. Prog. Phys. 47(1984) 925;
       K. Sato, in 'Cosmology of the Early Universe', ed. by L.Z. Fang and R. Ruffini
          (1984; World Scientific, Singapore) p.165;
       R.H. Brandenberger, Rev. Mod. Phys. 57(1985) 1; in 'Quantum Gravity and
          Cosmology', ed. by H. Sato and T. Inami (1986; World Scientific, Singapore)
          p.207;
       M.S. Turner, Fermilab preprint, FERMILAB-Conf-85/153 (1985; to be published
          in the Proceeding of the Cargese School on Fundamental Physics and Cosmology
          ed. by J. Audouze and J. Tran Thanh Van (Ed. Frontieres, Gif-Sur-Yvette)).

(6) K. Maeda, Class. and Quantum Grav. 3(1986) 233; 3(1986) 651.

(7) K. Maeda and H. Nishino, Phys. Lett. 154B(1985) 358; 158B(1985) 381.

(8) C.G Callan, E.J. Martinec, M.J. Perry and D. Friedan, Nucl. Phys. B262(1986)
          593.

(9) A.H. Chamseddine, Nucl. Phys. B185(1981) 403;
       E. Bergshoeff, M. de Roo, B. de Wit and P. van Nieuwenhuizen, Nucl. Phys. B195
          (1982) 97;
       G.F. Chapline and N.S. Manton, Phys. Lett. 120B(1983) 105.

(10) B. Zwiebach, Phys. Lett. 156B(1985) 315;
       B. Zumino, Univ. of California, Berkeley, Report No. UCB-PTH-85/13 (to be
          published).

(11) K. Maeda, Phys. Lett. 166B(1986) 59.

(12) K. Maeda, ICTP preprint, IC/86/180 (1986).

(13) D. Bailin and A. Love, Phys. Lett. 163B(1985) 135;
       D. Bailin, A. Love and D. Wong, Phys. Lett. 165B(1985) 270.

(14) M. Dine, R. Rohm, N. Seiberg and E. Witten, Phys. Lett. 156B(1985) 55.

(15) E. Witten, Phys. Lett. 155B(1985) 151;
       J.-P. Derendinger, L. Ibanez and H. Nilles, Nucl. Phys. B267(1986) 365.

(16) R. Rohm and E. Witten, Ann. Phys. 170(1986) 454.

(17) K. Enqvist, D.V. Nanopoulos, E. Papantonopoulos and K. Tamvakis, Phys. Lett.
        166B(1986) 41.

(18) K. Maeda and P.Y.T Pang, SISSA preprint, 84/85/A (1985; to be published in
        Phys. Lett. B).

(19) S.H.H. Tye, Phys. Lett. 158B(1985) 388;
        B. Sundborg, preprint, Göteborg, 84-51 (1984);
        M.J. Bowick and L.C.R. Wijwardhana, Phys. Rev. Lett. 54(1985) 2485;
        M.J. Bowick, L. Smolin and L.C.R. Wijwardhana, Phys. Rev. Lett. 56(1986) 424;
        E. Alvarez, Phys. Rev. D31(1985) 418; Nucl. Phys. B269(1986) 596;
        M. Gleiser and J.G. Taylor, Phys. Lett. 164B(1985) 36.

(20) A. Guth, Phys. Rev. D23(1981) 347;
        K. Sato, Mon. Not. Roy. astron. Soc. 195(1981) 467;
        A. Albrecht and P.J. Steinhardt, Phys. Rev. Lett. 48(1982) 1220;
        A.D. Linde, Phys. Lett. 108B(1982) 389.

(21) A.A. Starobinski, Phys. Lett. 91B(1980) 99; Sov. Astro. Lett. 10(1984) 135.

(22) D. Sahdev, Phys. Lett. 137B(1984) 155; Phys. Rev. D30(1984) 2495;
        H. Ishihara, Prog. Theor. Phys. 72(1984) 376L;
        R.B. Abbott, S.M. Barr and S.D. Ellis, Phys. Rev. D30(1984) 720; D31(1985) 673;
        E.W. Kolb, D. Lindley and D. Seckel, Phys. Rev. D30(1984) 673;
        Q. Shafi and C. Wetterich, Phys. Lett. 129B(1983) 387; 152B(1985) 51;
        Y. Okada, Phys. Lett. 150B(1985) 103;
        D. Bailin, A. Love and J. Stein-Schabes, Nucl. Phys. B253(1985) 387.

(23) A.A. Tseytlin, Phys. Lett. 176B(1986) 92;
        S. Deser and A.N. Redlich, Phys. Lett. 176B(1986) 350.

(24) K. Maeda and M.D. Pollock, Phys. Lett. 173B(1986) 251.

(25) D.G. Boulware and S. Deser, Phys. Lett. 175B(1986) 409.

(26) K. Maeda, M.D. Pollock and C.E. Vayonakis, ICTP preprint, IC/86/5 (1986; to
        be published in Class. and Quantum Grav.);
        J. Ellis, K. Enqvist, D.V. Nanopoulos and M. Quiros, CERN preprint, CERN-TH.
        4325/85 (1985);
        P. Oh, Phys. Lett. 166B(1986) 292;
        M. Yoshimura, KEK preprint, KEK-TH 114 (1985).

MUTUALLY INTERACTING QUANTUM FIELDS IN CURVED SPACE-TIMES

Jürgen Audretsch
Fakultät für Physik
Universität Konstanz
Postfach 5560
D-7750 Konstanz
W.-Germany

I. INTRODUCTION

The discussion of the physical details of the Inflationary Universe
has demonstrated that it would be very useful to have a feasible scheme
with worked-out examples for <u>quantum-field theory of mutually (!) interacting fields in a given cosmological background space-time.</u> The sort of
questions one would like to answer are thereby of the type:

i.) What is an appropriate conceptual framework for such a theory ?
What are the measurable quantities ? What is a related calculation
scheme ?

ii.) What are the physical characteristics of the semi-classical situation, i.e. in which way does the gravitational background typically
influence quantum-field theoretical processes ? What is the structure of the underlying physics from a more qualitative point of
view ? What are appropriate concepts to reveal this ?

iii.) What are the physical and mathematical advantages and deficiencies
of a given scheme ? Does it reflect what one would like to describe
and to obtain from a more intuitive physical point of view ?

iv.) And, last but not least: How are the minkowskian cross sections and
decay rates modified at a certain early time during the cosmic evolution or at a certain place near a strongly gravitating source ?

Up to now we have no approach to describe such a localization of an
interaction process in an appropriate way. Only for the problem of back-
creation, localized information can be obtained from the expectation

values of currents like the stress tensor. For reviews see [1,2]. Further literature can be found in [3-6], compare also [7].

A practicable scheme for the discussion of mutual interaction is the S-matrix approach (in-out approach). The conceptual framework, the measurable quantities and the related calculation scheme have been discussed in detail in [3-6]. This approach is close to the procedure one is used to in flat space-time. It has the disadvantage of being non-local by nature if the background is non-minkowskian. On the other hand, when applying it, one knows rather clearly what one is conceptually doing. Therefore, the study of this approach is a useful preliminary stage at which the structure of the gravitational influence can still be worked out by means of exact calculations. On the basis of the physical insights gained thereby, physically justified approximations may later be found to solve the localization problem. Naturally, all qualitative results obtained with the in-out approach should be handled with care when using them as input for astrophysical calculations.

In the following we give a very brief summary of the results of our papers [3-6], where all details and exact calculations can be found. We discuss two main subjects: a.) Appropriately defined transition probabilities and their application to decay processes. b.) The effect of gravitationally induced amplification (and attenuation) and its demonstration by means of the Compton effect.

II. SEMI-CLASSICAL APPROACH

We study the interaction given by $\mathcal{L}_I(\lambda, g_{\alpha\beta}, \Phi, \Psi)$ between two types of neutral scalar particles described by the massive Klein-Gordon field Φ and the massless field Ψ. This may be regarded as a first step for the discussion of the scalar quantum electrodynamics. λ is the coupling parameter. The fact that the curved space-time background acts quantum-field theoretically already in zeroth order $\mathcal{L}_I = 0$ (for example in creating particles out of the vacuum), will thereby severely influence the outcome of the mutual interaction and its registration.

We restrict ourselves to a treatment in the interaction picture using an in-out scheme based on the S-matrix $S = \lim_{\alpha \to 0} \hat{T} \exp\{i \int \mathcal{L}_I e^{-\alpha|\eta|} d^4x\}$. The gravitational background is thereby always exactly taken into account. α is the switch-off parameter.

We consider a 3-flat Robertson-Walker Universe

$$ds^2 = a^2(\eta) \left(d\eta^2 - d\underline{x}^2\right) \tag{2.1}$$

which is conformally flat. Apart from the examples discussed below, the expansion law is left unspecified, but the in- and out-regions ($\eta = -\infty$, $\eta = +\infty$) must allow the definition of free particles.

The Klein-Gordon particles are assumed to be conformally coupled to the background ($\nabla_\mu \nabla^\mu + m + R/6) \Phi = 0$ (R = scalar curvature , for Ψ: $m=0$). The general results can easily be transcribed for other fields and other interactions.

The physical specification above has the following consequences: i.) energy is not conserved. ii.) there are conserved 3-momentum parameters called $\underline{p}$, $\underline{q}$, ... for Φ-fields and $\underline{k}$, $\underline{l}$ ... for Ψ-fields. iii.) there is creation of massive Φ-pairs out of the vacuum in every mode $\underline{p}$ with total number $N^{(0)}(\underline{p}^\Phi|0) = |\beta_{\underline{p}}|^2 = N^{(0)} (-\underline{p}^\Phi|0)$. iv.) Because of the conformal situation there is no corresponding creation of Ψ-particles $N^{(0)} (\underline{k}^\Psi|0) = = |\beta_{\underline{k}}|^2 = 0$.

The information regarding the influence of the gravitational background is essentially contained in the Bogoliubov coefficients relating in- and out-particle solutions of the Klein-Grodon equation: $\alpha_{\underline{p}} = = (u_{\underline{p}}^{in}, u_{\underline{p}}^{out})$, $\beta_{\underline{p}} = (u_{\underline{p}}^{in}, u_{-\underline{p}}^{out*})$ with $|\alpha_{\underline{p}}|^2 - |\beta_{\underline{p}}|^2 = 1$.

III. ADDED-UP TRANSISITION PROBABILITY

The curved background creates particles out of the vacuum. To construct nevertheless a physically and mathematically reasonable transition probability, we know that in our case, because of the conformal field equation, a massless particle registered in the out-region has not come out of the background, but has solely been created or influenced by the mutual interaction: massless Ψ-particles are good indicators, whereas massive Φ - particles are produced by the gravitational background, too. The new concept <u>added-up probability</u> is adapted to this situation:

$$w^{add} (s^\Psi | c^\Phi r^\Psi) = \sum_{all\ d} |<out\ d^\Phi s^\Psi | S | c^\Phi r^\Psi\ in>|^2 \qquad (3.1)$$

It answers the question: What is the probability that a particular state $|s^\Psi\ out>$ of massless particles will be found in the out-region, regardless of what has happened to the massive states ? It is important to note then this quantity can be obtained in working out in-in amplitudes (compare [3])

$$w^{add} (s^\Psi | c^\Phi r^\Psi) = \sum_{all\ d} |<\ in\ d^\Phi s^\Psi | S | c^\Phi r^\Psi\ in>|^2 \qquad (3.2)$$

This permits a <u>Feynman-diagram technique</u> which differs from the one in flat space-time only by the replacement of the plane waves for massive particles by appropriate exact solutions u_p^{in} of the field equation in curved space-time.

IV. EXAMPLE: DECAY OF A MASSIVE PARTICLE

The intention of the study of the following examples is to read off a rather complete survey of the structure of the gravitational influence in Robertson-Walker Universes independent of the type of expansion law $a(\eta)$ used. Because we have to have approximately flat in- and out-regions to make particle definitions possible, there are only two different types of monotonically increasing expansion laws: one with $a(\eta) \xrightarrow[\eta \to -\infty]{} a_i = const,$ $a(\eta) \xrightarrow[\eta \to \infty]{} \infty$ and the other with $a(\eta) \xrightarrow[\eta \to -\infty]{} a_i = const, a(\eta) \xrightarrow[\eta \to \infty]{} a_0 = const,$ where η is an appropriate time-parameter. The first has been discussed in [3]. The results will not be repeated here. The second type is called <u>statically-bounded</u> and will be discussed below (for details see [6]). In all cases one will expect that particles created gravitationally out of the background contribute to the process of mutual interaction. We show how the respective contributions can be read off from our results.

We calculate and discuss the physical quantities for the scale factor representing a <u>step</u> at $\eta = 0$:

$$\hat{a}(\eta) = \Theta(-\eta)\, a_i + \Theta(\eta)\, a_0 \qquad (4.1)$$

with $a_i = \sqrt{A-B}, a_0 = \sqrt{A+B}, A>B>0$. We then add the corresponding rigorous calculation of the <u>tanh-expansion law</u>

$$\tilde{a}(\eta) = \sqrt{A+B \tanh b\eta} \ , \ b > 0 \tag{4.2}$$

which is as smoothed-out step the prototype of a statically bounded expansion law, leading to all the typical physical deviations from the step situation. Quantities with hat and tilda will refer to the respective expansion laws above.

In the following we sketch the exact calculation of the decay of a massive scalar Φ-particle into two massless scalar Ψ-particles in lowest order of the $(-\lambda/a(\eta)) \ \Phi\Psi^2$- interaction with coupling parameter λ. The factor a^{-1} makes the interaction conformally invariant, so that the exact calculation below becomes less cumbersome. The mass m in the field equations breaks the conformal invariance.

<u>Figure 1:</u>

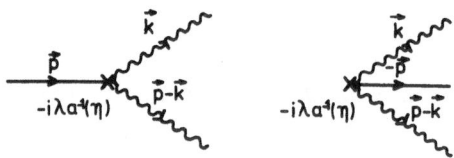

Diagrams contributing to the added-up transition probability

To derive the added-up probability for the decay process, we have to work out the probability amplitudes related to the two diagrams of figure 1. We obtain in the step case

$$\hat{w}^{add} = \frac{\lambda^2 \pi}{2k|\underline{p}-\underline{k}|V} \left[\frac{1}{2E_i} T_\eta \ \delta(\omega_{-i}) + (\frac{1}{2} + |\hat{\beta}_{\underline{p}}|^2) \frac{1}{E_o} T_\eta \ \delta(\omega_{-o}) \right] + \hat{\Delta}(k) \tag{4.3}$$

and in the tanh case

$$\tilde{w}^{add} = \frac{\lambda^2 \pi}{2k|\underline{p}-\underline{k}|V} \left[\frac{1}{2E_i} (T_\eta + \text{finite}) \ \delta(\omega_{-i}) + \right.$$

$$\left. + (\frac{1}{2} + |\tilde{\beta}_{\underline{p}}|^2) \frac{1}{E_o} (T_\eta + \text{finite}) \ \delta(\omega_{-o}) \right] + \tilde{\Delta}(k) \tag{4.4}$$

$E_{i/o} = \sqrt{p^2 + m^2 a_{i/o}^2}$ is thereby the energy parameter of the massive particles. With $\omega_{-i/o} = E_{i/o} - (k_1 + k_2)$ the conservation of the measured (!) energy in the in-out-region is given by $\omega_{-i/o} = 0$. We have introduced the infinite η-time T_η between the in- and out-regions during which the quantum interaction takes place: $2\pi\delta(o) = \lim_{T_\eta \to \infty} T_\eta$. The typical spectral behaviour of the $\tilde{\Delta}(k)$ is shown in figure 2. It shows smoothed-out resonances around energy conservation in the in-region ($k = E_o/2$) and in the out-region ($k = E_o/2$).

Figure 2:

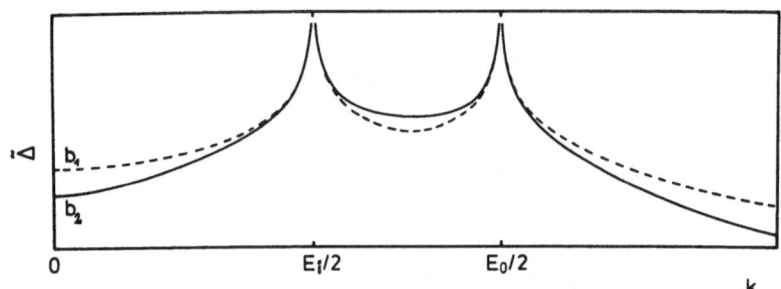

The correction $\tilde{\Delta}(k)$ in (4.4) as a function of the energy of the massless particles for different values of the expansion parameter b for decaying particles at rest ($\underline{p} = 0$).

Apart from the fact that w^{add} is in both cases, because of the Δ-terms, non-vanishing outside of the resonances, we see that smoothing-out the step leads formally to _additive bounded_ corrections of the infinite interaction time T_η. Furthermore it is very typical that the out-resonance is amplified by a factor $(1+2|\beta_p|^2)$ as compared with the in-resonance. This factor contains the mean number of massive particles $|\beta_p|^2$ gravitationally created out of the background in zeroth order of the mutual interaction. An explanation will be given below.

We turn to a discussion of the origin and the physical significance of the two different resonances in w^{add} and study the step-expansion law first. In the calculations given above, massive and massless particles are identified not by measured quantities, but by the momentum parameters $\underline{p}$ and $\underline{k}$, respectively. On the other hand, the decay of a massive particle into two massless ones will become resonant if conservation of 3-momentum as well as conservation of _measured_ energy is fulfilled. The latter means $\omega_{-i} = 0$ for $\eta \leq 0$, which can be written as $\sqrt{\underline{p}^2 + m^2 a_i^2} = k_1 + k_2$. For $\eta \geq 0$ we have $\dot{\omega}_{-o} = 0$ and therefore $\sqrt{\underline{p}^2 + m^2 a_o^2} = k_1 + k_2$. Accordingly, for ingoing Φ-particles prepared with a given momentum $\underline{p}$, decay and energy conservation happening either in the region $\eta \leq 0$ or in the region $\eta \geq 0$, implies for these two cases __different__ values of the parameters k_1 and k_2 of the outgoing Ψ-particles. But in any case the measurement of the Ψ-particles is performed in the out-region. This causes the appearance of __two resonances__ in the spectrum of w^{add}, which are typical for statically bounded expansion laws.

For the __total added-up probability,__ $w^{tot} = \sum_{\underline{k}} w^{add} (1^{\Psi}_{\underline{k}} 1^{\Psi}_{\underline{p}-\underline{k}} | 1^{\Phi}_{\underline{p}})$, we obtain in the case $\underline{p} = 0$

$$\hat{w}^{tot} = \frac{\lambda}{4\pi m} \left[\frac{1}{2 a_i} T_{\eta} + \frac{1}{a_o} (\frac{1}{2} + |\hat{\beta}_{\underline{p}=o}|^2) T_{\eta} \right] \tag{4.5}$$

$$\tilde{w}^{tot} = \frac{\lambda}{4\pi m} \left[\frac{1}{2a_i} T_{\eta} + \frac{1}{a_o} (\frac{1}{2} + |\tilde{\beta}_{\underline{p}=o}|^2) T_{\eta} \right] + R^{fin} (a_i, a_o, |\tilde{\beta}|^2) \tag{4.6}$$

R^{fin} is thereby a finite additive correction.

We give an interpretation of the two w^{tot} in discussing the question: How is the result (4.5) for the step-law related to the minkowskian total transition probability $w^{tot}_{Mink} = (\lambda^2/4\pi m) T_t$? In our case one half of the particles have a chance to decay in the Minkowski-region $\eta \leq 0$ with $a(\eta) = a_i$ (and the others in the region $\eta \geq 0$ with $a(\eta) = a_o$). Furthermore, because of the structure of $\mathcal{L}_I$, the interaction contains for $\eta \leq 0$ a factor λa_i^{-1} instead of λ as in the minkowskian case ($a = 1$). Introducing finally according to $T_t = a_i T_{\eta}$ the conformal time T_{η}, we end up with the first term of (4.5) The second is obtained by corresponding considerations related to the interval $\eta \geq 0$.

With regard to the third term of (4.5) we recall that $|\hat{\beta}_{\underline{p}=o}|^2$ is the number of massive particles per unit volume and the momentum interval around $\underline{p} = 0$ which are gravitationally created out of the background in the zeroth order of the mutual interaction. The appearance of this third term reflects the fact that not only one half of the incoming particles , but in addition also these created particles are decaying in the region $\eta \geq 0$. Therefore, as compared with the second term and its underlying process, the factor 1/2 has, in this case, to be replaced by $|\hat{\beta}_{\underline{p}=o}|^2$ while the factor $1/E_o$ remains unchanged.

The three resonant terms proportional to T_η in $\tilde{w}^{tot}$ of (4.6) also go back to the minkowski-type processes. But instead of taking the mean value of the two minkowski-type contributions we have to work out the temporal mean value of the outcome of the processes happening in the continuous family of tangent-spaces. To do so, we make use of $<\frac{1}{a(\eta)}> = \frac{1}{2}\left[\frac{1}{a_i} + \frac{1}{a_0}\right]$. Because this relation is true for all statically bounded monotonic expansion laws, we find that also in the general case the resonant terms, i.e. the terms proportional to T_η, agree with those in $\hat{w}^{tot}$, provided we repalce the Bogoliubov coefficients in the usual way.

The appearance of the additional finite term R^{fin} in $\tilde{w}^{tot}$ is the important generic consequence of the fact, that the gravitational influence of the interaction process is only of finite duration. There is no such term in $\hat{w}^{tot}$ because the gravitational influence happens only at the one point of time $\eta = 0$.

The transition probabilities (4.5) and (4.6) still contain the infinite duration T_η of the mutual interaction. In Minkowski space-time ($a(\eta) = 1$) the usual procedure would be to divide the related probabilities by the time $T_\eta = T_t$, thus obtaining as a physically relevant quantity the reciprocal lifetime of the massive particles at rest $\tau^{-1}_{Mink} = \lambda^2/4\pi m$.

Referring in the following to the η-time, the same procedure can be applied in the step-case (4.5) leading to

$$\frac{1}{\hat{\tau}} = \frac{1}{\tau_{<Mink>}} + \frac{\lambda^2}{4\pi E_0}\left|\hat{\beta}_{\underline{p}=0}\right|^2 \tag{4.7}$$

with $\tau^{-1}_{<Mink>} = \left(\frac{\lambda^2}{4\pi}\right)\left(\frac{1}{2E_i} - \frac{1}{2E_0}\right)$. The latter represents the reciprocal η-lifetime going back to the temporal mean value of the local minkowskian contributions as described in the preceding paragraph. Because we had to base our discussion on the concept of the added-up transition probability (3.1-2), the decay product of those Φ-particles which are created out of the background enter the calculation of $\hat{\tau}$. It is possible to eliminate this influence in omitting the $|\hat{\beta}_{\underline{p}=0}|^2$-term in (4.7).

On the other hand, to work out the η-lifetime $\tilde{\tau}$ in the tanh-case or in other non-step-cases, we have to draw attention to the fact, that in

these cases the duration of the mutual interaction on one hand, and the
gravitational influence caused by the curved background on the other, are
characterized by two different time scales. The first time scale is, because of the adiabatic switch-off in S , again the infinite time T_η. In
contrast to this, the second time scale which represents the duration
of the gravitational influence on the mutual interaction is finite. It
will be called the gravitational time T^{grav}. The appearance of these two
time scales seems to be a characteristical trait of an S-matrix approach
in a given curved space-time: In order to be able to introduce particles
in the asymptotic in- and out-regions, the curved part of the space-time
must be localized, thus implying the second finite time scale.

We have to stress, however, that from the point of view of the exact-
ness there seems to be no unambiguous and generally applicable way to define such a T^{grav} rigorously. Nevertheless, to obtain some quantitative
ideas about the gravitational influence of the decay process, we may
divide - according to their respective origin - the divergent part of
$\tilde{w}^{tot}$ by the infinite time T_η and the finite rest R^{fin} by T^{grav} to
obtain:

$$\frac{1}{\tilde{\tau}} = \frac{1}{\tau_{<Mink>}} + \frac{\lambda^2}{4\pi E_0} |\tilde{\beta}_{p=0}|^2 + \frac{R^{fin}}{T^{grav}} \tag{4.8}$$

V. GRAVITATIONALLY INDUCED AMPLIFICATION AND ATTENUATION

We turn now in the second part of this paper to another important
effect which governs quantum field theory in given curved space-times.
The mean number of outgoing massive Φ-particles in the mode $\underline{p}$ is

$$N(\underline{p}^\Phi|a) = \sum_{all\ b} |<out\ b|S|a\ in>|^2\ n(\underline{p}^\Phi|b) \tag{5.1}$$

if the in-state was $|a\ in>$. Parker [8,9] has shown that the respective
zeroth order expression has the structure

$$N^{(0)}(\underline{p}^\Phi|a) = N^{(0)}(\underline{p}^\Phi|0) + n(\underline{p}^\Phi|a) + N^{(0)}(\underline{p}^\Phi|0)\left[n(\underline{p}^\Phi|a) + n(-\underline{p}^\Phi|a)\right] \tag{5.2}$$

where $n(\underline{p}^\Phi|a)$ is the number of Φ-particles occupying the $\underline{p}$-mode of $|a>$.
The meaning of the three terms in (5.1) is: particle creation out of the

vacuum, particles which have passed through and, finally, <u>gravitationally</u> <u>induced amplification</u> of the ingoing particle content. This amplification results in additional outgoing pairs. The latter fact is indicated by the appearance of $n(-\underline{p}^{\Phi}|a)$, according to which ingoing particles in the mode $-p$ induce creation in the mode $\underline{p}$.

Fermions, on the other had, show <u>attenuation</u> (negative third term). For complex fields the $-\underline{p}$-mode is an antiparticle mode.

For higher orders of the mutual interaction we obtain correspondingly (for details see [4])

$$N^{(z)}(\underline{p}^{\Phi}|a) = \sum_{\text{all b}} \left| <\text{in } b|S|a \text{ in}> \right|^2_{(z)} n(\underline{p}^{\Phi}|b) +$$

$$+ N^{(o)}(\underline{p}^{\Phi}|0) \sum_{\text{all b}} \left| <\text{in } b|S|a \text{ in}> \right|^2_{(z)} [n(\underline{p}^{\Phi}|b) + n(-\underline{p}^{\Phi}|b)] + \qquad (5.3)$$

$$+ \text{Re}(\beta_{\underline{p}}^* \sigma_{\underline{p}})$$

with

$$\sigma_{\underline{p}} = -2\alpha_{\underline{p}}^* \sum_{\text{all b}} <\text{in } a|S^{\dagger}|b \text{ in}> <1_{\underline{p}}^{\Phi} 1_{-\underline{p}}^{\Phi} b \text{ in}|S|a \text{ in}> \qquad (5.4)$$

The second term is again the <u>amplification</u>, now being a part of the mutual interaction. The third term has no correspondence in the zeroth order formula.

The structure of (5.4) may be visualized using the following diagrammatic rule: "Let the mutual interaction happen completely within the in-region only and describe it accordingly by in-in transition amplitudes. Process now the corresponding particle outcome as in zeroth order in a twofold way: At first the particles in the p-mode pass through into the out-region as in (5.2) to obtain the first term of (5.3). Secondly, these particles are amplified in the same way as in (5.2) whereby a possible outcome in the $-p$-mode contributes in a symmetric manner. This leads to the second term in (5.3). Finally the σ-term of (5.3) is to be added". When using this rule it must not be forgotten that the calculation is in fact based on one single coherent in-out process.

Writing the particle number according to

$$N^{(z)}(\underline{p}^{\Phi}|a) = \sum_{\text{all } b} |<\text{in } b|S|a \text{ in}>|^2_{(z)} \, n(\underline{p}^{\Phi}|b)\{1 + N^{(o)}(\underline{p}^{\Phi}|0)\left[1 + \frac{n(-\underline{p}^{\Phi}|b)}{n(\underline{p}^{\Phi}|b)}\right]\}$$

(5.5)

$$+ \, \text{Re}(\beta_{\underline{p}}^* \sigma_{\underline{p}})$$

we can read off that gravitational amplification acts as a mode-dependent amplification factor (!) and not as an additive term. This means that the minkowskian contributions contained in the in-in transition amplitude are altered in a multiplicative way, which may lead to considerable modifica-- tions.

Fermions show for non-zero order of the mutual interaction again attenuation instead of amplification. See the appendix of reference [4] for details.

An immediate consequence of (5.2) and (5.3) is

$$N(\underline{p}^{\Phi}|a) - N(-\underline{p}^{\Phi}|a) = \sum_{\text{all } b} |<\text{in } b|S|a \text{ in}>|^2[n(\underline{p}^{\Phi}|b) - n(-\underline{p}^{\Phi}|b)]$$

(5.6)

Asymmetry in the particle content of the outgoing p- and -p-mode can therefore be solely caused by the structure of the mutual interaction as represented by S. Amplification and the process leading to the σ- term in (5.3) all happen as creation of (p, -p)-pairs. Thus reflecting 3-momentum conservation. In the charged case we would find particle- and antiparticle pairs.

In reference [4] we have studied the example of particle creation out of the vacuum for the interaction $\mathcal{L}_I = -\sqrt{-g}(\lambda/a^2(\eta))\Phi\Psi$ and the expansion law $a^2(\eta) = 1 + e^{2b\eta}$. This has been done exactly up to the order λ^2. The multiplicative amplification and the role of the σ-term are discussed in detail.

VI. IMPROVED INDICATOR CONFIGURATIONS

We can improve the predictive power of the concepts introduced above in specifying the massive part of the end state too and in using less extensive summations as compared with (3.1) and (5.1). Good massive indi-

<u>cators</u> are all configurations without massive particle pairs, because they consist of massive particles which originate from the interaction only. The corresponding probability that such a transition has occurred regardless of the creation of massive pairs out of the background or interaction is the <u>pair-including transition probability.</u>

$$w^{inc}(\hat{d}^{\Phi}s^{\Psi}|c^{\Phi}r^{\Psi}) = \sum_{all \; Q} |<out \; Q^{\Phi}\hat{d}^{\Phi}s^{\Psi}|S|c^{\Phi}r^{\Psi} \; in>|^2 \qquad (6.1)$$

The sum thereby goes over all possible states Q which consist of massive pairs. Such pair states are indicated by capital letters.

w^{inc} can again be built up out of in-in amplitudes (compare [5])

$$w^{inc}(\hat{d}^{\Phi}s^{\Psi}|c^{\Phi}r^{\Psi}) = \sum_{all \; Q} |< in \; Q^{\Phi}\hat{d}^{\Phi}s^{\Psi}|S|c^{\Phi}r^{\Psi} \; in >|^2 \qquad (6.2)$$

This guarantees that for any finite order of the mutual interaction the sum over Q stops, ending with states Q which contain only a particular finite number of massive particles. A perturbation scheme based on Feynman rules which are again as simple as in flat space-time may therefore be established. In the framework sketched above, this probability w^{inc} seems to be a concept as close as it can be to what we are used to in flat space-time.

Specifying again not only the in- but also the unpaired part of the out-state and allowing for the production of pairs as above (thus isolating the particular transition process as far as possible) we are led to the following concept of a <u>specified mean number</u> N($\leftarrow$) :

$$N(\underline{p}^{\Phi}|\hat{d}^{\Phi}s^{\Psi} \leftarrow c^{\Phi}r^{\Psi}) = \sum_{all \; Q} |< out \; Q^{\Phi}\hat{d}^{\Phi}s^{\Psi}|S|c^{\Phi}r^{\Psi} \; in>|^2 \; n(\underline{p}^{\Phi}|Q^{\Phi}\hat{d}^{\Phi}s^{\Psi}) \qquad (6.3)$$

Subtracting the contribution of particles which originate from vacuum-creation leads to $N^{int}(\leftarrow)$ which refers to the mutual interaction only, and can be transcribed into (compare [5])

$$N^{int}(\underline{p}^{\Phi}|\hat{d}^{\Phi}s^{\Psi} \leftarrow c^{\Phi}r^{\Psi}) = \sum_{all \; Q} |< in \; Q^{\Phi}\hat{d}^{\Phi}s^{\Psi}|S|c^{\Phi}r^{\Psi} \; in>|^2 n(\underline{p}^{\Phi}|Q^{\Phi}\hat{d}^{\Phi}s^{\Psi}) +$$

$$+ N^{(0)}(\underline{p}^{\Phi}(0) \sum_{all \; Q} |< in \; Q^{\Phi}\hat{d}^{\Phi}s^{\Psi}|S|c^{\Phi}r^{\Psi} \; in>|^2 \left[n(\underline{p}^{\Phi}|Q^{\Phi}\hat{d}^{\Phi}s^{\Psi}) + n(-\underline{p}^{\Phi}|Q^{\Phi}\hat{d}^{\Phi}s^{\Psi}) \right] + \qquad (6.4)$$

$$+ Re(\beta_{\underline{p}}^{*} \tilde{\sigma}_{\underline{p}})$$

with

$$\tilde{\sigma}_{\underline{p}} = -2\,\alpha_{\underline{p}}^{*} \sum_{\text{all } L} <\text{in } c^{\Phi}r^{\Psi}|S^{\dagger}|L^{\Phi}\hat{d}^{\Phi}s^{\Psi} \text{ in}><\text{in } L^{\Phi}1_{\underline{p}}^{\Phi}1_{-\underline{p}}^{\Phi}\hat{d}^{\Phi}s^{\Psi}|S|c^{\Phi}r^{\Psi} \text{ in} > \tag{6.5}$$

Summation over $\hat{d}$ reproduces (5.2) and (5.3). The first term, therefore, is a weighted particle creation out of the vacuum. The second term is again the amplification which shows its specific structure already on this level.

VII. EXAMPLE: COMPTON-EFFECT IN THE $\Phi^2\Psi^2$-MODEL REFLECTS GRAVITATIONALLY INDUCED AMPLIFICATION

We study the Compton scattering in the interaction $\mathcal{L}_I = -\sqrt{-g}\,\lambda\,\Phi^2\Psi^2$ outside of forward scattering. We disregard the contribution resulting from pair creation out of the vacuum and concentrate on the specified mean numbers which refer to the mutual interaction only $N^{\text{int}}(\leftarrow)$. This agrees because of the conformal coupling with $N(\leftarrow)$ in the massless Ψ-case. Discussion of the amplitudes in (6.4) then leads directly to

$$N^{\text{int}}(\underline{p}^{\Phi}|1_{\underline{p}}^{\Phi}1_{\underline{k}}^{\Psi} \leftarrow 1_{\underline{q}}^{\Phi}\,1_{\underline{1}}^{\Psi}) = N^{\text{int}}(\underline{k}^{\Phi}|1_{\underline{p}}^{\Phi}1_{\underline{k}}^{\Psi} \leftarrow 1_{\underline{q}}^{\Phi}1_{\underline{1}}^{\Psi})\Big[1+N^{(0)}(\underline{p}^{\Phi}|0)\Big| + 0(\lambda^3) \tag{7.1}$$

where $1_{\underline{p}}^{\Phi}\,1_{\underline{k}}^{\Psi}$ is the end-state and $1_{\underline{q}}^{\Phi}\,1_{\underline{1}}^{\Psi}$ is the initial state of Compton scattering, as has been shown in [5].   (7.1) clearly demonstrates for the Compton effect the gravitationally induced amplification: Massive Φ-particles and massless Ψ-particles leaving the mutual interaction are not going out in equal number, as one would expect from the situation in flat space-time or from the Feynman diagram. Rather the number of massive particles is amplified by a momentum-dependent factor

$$\Big[1 + N^{(0)}(\underline{p}^{\Phi}|0)\Big] = \Big[1 + |\beta_{\underline{p}}|^2\Big]$$

REFERENCES

1) N.D. Birrell, in: Quantum Gravity 2, eds. C.J. Isham, R. Penrose and D.W. Sciama (Oxford University Press, Oxford, 1981).

2) L. Ford, in: Quantum theory of gravity, ed. Christensen (Adam Hilger, Bristol, 1984).

3) J. Audretsch and P. Spangehl, Class. Quantum Grav. 2 (1985), 733.

4) J. Audretsch and P. Spangehl, Phys.Rev. D 33 (1986), 997

5) J. Audretsch and P. Spangehl, Improved concepts for the discussion of mutually interacting quantum fields in Robertson-Walker universes, preprint University of Konstanz (1985).

6) J. Audretsch, A. Rüger and P. Spangehl, Decay of massive particles in Robertson-Walker universes with statically bounded expansion laws, preprint University of Konstanz (1986).

7) L. Ford, Phys.Rev. D31 (1985), 704.

8) L. Parker, Phys.Rev. 183 (1969), 1057.

9) L. Parker, Phys.Rev. D3 (1971), 346.

GRAVITONS IN DE SITTER SPACE

Bruce ALLEN
Department of Physics and Astronomy
Tufts University
Medford, MA 02155
U.S.A.

Everyone knows that "Einstein's greatest mistake", the cosmological constant Λ, is very close to zero |1|. There have been many attempts to explain why Λ must be exactly zero |2|, but none of these efforts has succeeded. In fact it is now fashionable to believe that during the very early history of the universe the value of Λ was quite large |3|. This so-called "inflationary" epoch would have been a long period of exponentially rapid expansion, and would elegantly explain two otherwise mysterious observational truths : the universe is uncannily flat, and the cosmic microwave background radiation has no right to be as isotropic as it is.

There are three things that we would like to know. First, why is Λ zero today ? Second, could Λ have been very big in the past ? And finally, if Λ was very big in the past, what consequences would that have today ? Unfortunately this paper will not answer any of these questions, but I hope that it will nevertheless accomplish something useful. I am going to show that one of the answers that has recently been given to the first question above - Why is Λ zero ? - is not correct. However, before I get into the technical nitty-gritty, let me give you a synopsis of the kinds of answers that have been suggested to these questions.

One answer which has been given to the question - why is Λ equal to zero today ?- has been that zero is the only consistent answer. Let me reveal my predjudices at once and say that I don't believe this. First of all, it isn't borne out by careful calculation. For example, someone once decided that Λ must be zero, because if it was not zero then a certain scattering amplitude would not be unitary. However a more careful investigation showed this to be false |4|. A different argument, which is being worked on by several people, is that if Λ is not zero then particles get created out of the vacuum, and damp the value of Λ to zero |5|.

The problem is this : What quantities to you calculate to see if Λ really decays (or has to be zero from the outset) ? One may show, for example, that a scalar particle propagating in a background with Λ nonzero will emit additional scalar particles, which will continue to do the same thing, and so on, ad infinitum. Now this <u>sounds</u>

unstable. However if you calculate the energy-momentum tensor of this process, you find that it only shifts the value of Λ a little bit |6|.

Another example : you can calculate the corrections to the stress-tensor due to the Plank-scale quantum fluctuations of the vacuum. Indeed, these corrections shift the value of Λ , and one can study the semi-classical back-reaction to find out what effect this has on the metric tensor. Do this, and you find that the metric is greatly altered. It sounds like a tremendous physical instability until you realize that the identical argument implies that flat space with Λ = 0 can't be stable either ! |7|. So here it is clear that something is wrong with the argument itself, since locally our spacetime is very flat, and shows no signs of decaying away beneath our feet !

One of the basic problems with these arguments is that the natural ground state (or vacuum state) for de Sitter space is time-reversal symmetric |8|. In this so-called Gibbons-Hawking vacuum state, it is impossible for particle creation to occur, because if the number of particles were increasing, that would break time-reversal symmetry.

If it could really be established that Λ = 0 was the only consistent value, I don't think that it would be a good thing . The cosmological constant is a measure of the local energy density of the empty vacuum state. If it was truly zero then there would be no way to generate cosmological inflation, which would be very unfortunate. (This is another reason why I am inclined to believe that there is nothing which is inconsistent about Λ nonzero). Because Λ is simply a measure of the vacuum or latent energy, it can change during phase transitions, and it seems certain that if symmetry in gauge theories is restored at high temperatures then such phase transitions must have taken place, as our universe cooled and expanded |9|. So it seems quite possible that Λ was nonzero in the past-and this leads to our final question above.

What kinds of effects would be associated with a large positive Λ ? Well, first there would be classical "gravitational" effects. Separated particles, freely falling, would accelerate away from one another. At a certain distance the recessional velocity would become unity, and there would be a cosmological particle horizon |10|. A given observer could not see farther than this distance. In addition to these classical effects there would be quantum effects. The best known of these is the Gibbons-Hawking effect -a freely falling observer would see a thermal spectrum at temperature $(12\pi^2)^{-1/2}\Lambda^{1/2}$ radiating from the imaginary surface which we have just described-the observer's particle horizon |8|. There are probably other interesting effects too, but we don't know what they are yet. The subject of quantum field theory in de Sitter space is still in its infancy. In fact the only real results are that we know how to construct the Fock space of states, and how to find the correlation functions for spin 0, ½ and 1 |11|. We also know a little bit about interacting fields in de Sitter space |12|.

Now let me tell you the idea which I intend to spend the rest of this talk trying to destroy. The idea was to show that Λ must be zero because of properties of gravitons when Λ is nonzero. If we were trying to prove that $\Lambda = 0$, then, in the absence of any exact symmetry or invariance which would force Λ to vanish, this would be the next best thing. The reason is this : Λ is only observable through relativity, because it represents an otherwise completely arbitrary zero-point for measuring the energy-density of space-time. Were it not for general relativity, or gravity, then any background energy density would be completely unobservable, and it could be set to zero with the stroke of a pen. However in the presence of gravity, the vacuum energy density Λ does become observable, for example through the classical effects described above. It would therefore be nice if the only thing (gravity) that enables us to observe Λ in the first place would also carry with it some quantization consistency condition that would force Λ to be zero. This would be an elegant solution to our problem : if gravity, the only force that allows us to observe Λ, demands that Λ vanish for reasons of consistency.

An argument of this type has recently been made by Antoniadis, Iliopoulos and Tomaras |13|. They claim that if one quantizes gravitational fluctuations in the presence of a background energy density Λ, then the resulting theory shows a particular kind of inconsistency called an "infra-red divergence". Now these words can refer to any one of several different problems. For example saying that QED has infra-red divergences generally refers to the fact that external lines in Feynman diagrams emit an infinite number of low-frequency photons |14|. However this is not a real problem because the energy carried away by this process is not infinite. Similarly, in the theory of massless scalar electro-dynamics, the effective action has an infra-red divergence, and a mass-scale must therefore be introduced into the theory. The gauge fields thus aquire a mass, and the gauge symmetry is broken |15|.

In the recent work by Antoniadis, Iliopoulos and Tomaras |13|, it was claimed that, because of an infra-red divergence, the two-point function of gravitational fluctuations was infinite (regardless of the separation of the two points). They argued that this infinity caused certain tree-level scattering amplitudes to be infinite, and thus rendered the theory of quantum gravity inconsistent unless Λ equaled zero.

What I am going to do in this talk is quite straightforward. First, I am going to talk about the graviton propagator, and explain why it is not, in and of itself, a physical object. In fact it depends upon the choice of gauge (by which, as I will shortly explain, I mean the choice of a gauge-fixing term). What this means in practice is that physical quantities (for example scattering amplitudes, or the expectation value of the curvature tensor) depend only upon certain components of the propagator. This dependence is just subtle enough so that the different graviton propagators, arising from different choices of gauge, give exactly the same physical result.

The next thing that I will do is to show that it is indeed true that for certain choices of gauge, the graviton propagator is indeed infinite, exactly as claimed by Antoniadis et al. However I am then going to show that there are <u>other</u> choices of gau-

ge for which the propagator is completely finite ! Then I will explain why this is so.
The point will be that the gauge-fixing terms of Antoniadis et al. do not completely
fix the gauge because they still allow a finite number of gauge transformations. It
is for this reason that the propagator that they find is infinite. But this infinity
is not a real physical divergence ; it is an artifact of how the gauge-fixing was
done. For a better choice of gauge, the propagator is completely finite ! To make this
point absolutely clear, I will then show that if one calculates scattering amplitudes
in the Antoniadis et al. gauge, one still obtains a perfectly finite result, in spite
of the fact that the graviton propagator is infinite. The reason for this is that the
graviton propagator in their gauge is the sum of an infinite (unphysical gauge-arti-
fact) term and a finite part. The infinite term does not contribute to the scattering
amplitude of any interaction whose stress-tensor is conserved, and thus physical scattering amplitudes remain completely finite, contradicting the claims of Antoniadis et
al. Although I will not show it here, this cancellation takes place at higher orders
as well. The point is that for good choices of gauge, the Fadeev-Popov ghosts are well-
behaved ; for a bad choice of gauge they introduce additional infra-red divergences
in just the right way to cancel those arising from the gravitons.

So the real point of all this technical investigation is that, at least so far,
there doesn't seem to be any intrinsic problem with $\Lambda \neq 0$. Of course it's entirely
possible that something else will turn up in the future that will render de Sitter
space inconsistent ; as things stand at the moment, it seems that we have to keep on
thinking about it.

I. Gauge-Fixing Terms and the Choice of Gauge.

This is a straightforward subject, but one that a great many people seem to be unclear about. The source of most of the trouble is confusion about the relationship
between the classical process called "choice of gauge" and its analogue in quantum
field theory, which is called "choice of a gauge-fixing term" in the action. Let us
begin by considering these two ideas, and the connections between them.

Suppose that h_{ab} is a small perturbation of some background metric g_{ab}. Then there
is a whole class of metric perturbations h_{ab} that represent exactly the same physical
perturbation. This is because under the infinitesimal coordinate transformation
$x^i \rightarrow x^i + v^i$ the metric perturbation transforms into $h_{ab} + \nabla_{(a} v_{b)}$. Since coordinate
transformation does not cause any changes to physically observable quantities, we can
conclude that the perturbations h_{ab} and $h_{ab} + \nabla_{(a} v_{b)}$ are gauge-equivalent. Any physical quantity, for example the perturbation of the curvature induced by $h_{ab} + \nabla_{(a} v_{b)}$,
will not depend upon v_b |16|.

For this reason, in classical perturbation and stability theory, it is very common to "impose gauge conditions". The metric perturbations obviously lie in equivalence classes ; two perturbations will be deemed equivalent iff they differ by $\nabla_{(a}V_{b)}$ for some vector V_b. "Imposing gauge conditions" is a way to pick out one particular member from each equivalence class. For example one can impose the following conditions on h_{ab},

$$
\begin{aligned}
\nabla^a h_{ab} &= 0 && \text{(transverse)}, \\
h^a_a &= 0 && \text{(traceless)}, \\
t^a h_{ab} &= 0 && \text{(synchronous)},
\end{aligned}
\tag{1.1}
$$

to restrict the gauge freedom. Here t^a is some arbitrary vector field (usually choosen to be timelike). These are not the only conditions that one could impose ; there are clearly an infinite number of other possibilities.

Now what about the quantum field theory of gravitational perturbations ? Well the action is a scalar and it is thus invariant under coordinate transformations, so the perturbations h_{ab} and $h_{ab} + \nabla_{(a}V_{b)}$ have exactly the same action ; they are gauge equivalent. In this situation the standard thing to do is to add to the action an arbitrarily choosen term which is <u>not</u> invariant under the above transformation. This arbitrarily choosen gauge-fixing term breaks the gauge invariance. For example it could be,

$$
S_{gauge} = \int \left[\alpha(\nabla^a h_{ab})^2 + \beta(h^a_a)^2 + \gamma(t^a h_{ab})^2 \right] d(\text{Vol}),
\tag{1.2}
$$

where at least one of the positive constants (α, β, γ) was nonzero. Now of course we have made a very arbitrary choice here ; the Fadeev-Popov procedure allows us to compensate for this choice in just such a way that the scattering amplitudes are ultimately independent of it.

Now suppose that we have determined the euclidean propagator, which we could write for example as the path integral |17|

$$
G_{abc'd'}(X,X') = \int d[h_{ab}] h_{ab}(X) h_{c'd'}(X') \exp(-S[h_{ab}] - S_{gauge}[h_{ab}]).
\tag{1.3}
$$

This propagator obviously depends upon the gauge-fixing parameters α, β and γ. Now suppose that we considered the divergence $\nabla^a G_{abc'd'}$ as a function of α, β and γ. In general it would not vanish. However if α was taken to infinity, then it would vanish. The reason is that if α is very large then the field configurations in the action which don't have $\nabla_a h^{ab} = 0$ are exponentially suppressed by the gauge-fixing term. In the limit $\alpha \to \infty$ such configurations would make no contribution to the propagator. Similarly, if α, β and γ were simultaneously taken to infinity, then the propagator would satisfy the "classical gauge conditions" (1.1) in that $\nabla^a G_{abc'd'}$,

$G^a_{ac'd'}$, and $t^a G_{abc'd'}$ would all vanish. So we can see that the classical gauge conditions are obtained in quantum theory by singular choices of gauge. This is analogous to the situation in QED. There, if we wanted to have a transverse propagator satisfying $\nabla_a \langle A^a A^{b'} \rangle = 0$ we would need to choose Landau gauge ; ie use a gauge-fixing term $\lambda (\nabla_\mu A^\mu)^2$ in the action and take the limit $\lambda \to \infty$.

Now what gauge was used by Antoniadis et al. |13| ? In fact there were two possibilities that they considered. The first one was equivalent to the three conditions given before (1.1) where t^a is a vector field orthogonal to a family of flat spatial surfacer which are the standard k = 0 foliation of de Sitter space. In this choice of gauge the propagator for gravitons can be related to that of a pair of minimally coupled scalar fields in a simple manner. Indeed in this gauge (corresponding to taking α, β and γ to infinity in (1.2)) the propagator is infinite. However it was not clear if the reason for this was because the three gauge-fixing parameters were becoming infinite, or if it was because the introduction of the vector field t^a into the action was breaking de Sitter invariance, or if it was because the gauge-conditions did not entirely determine the gauge. At that point in their work, Antoniadis et al. were not themselves certain if the infra-red divergence that they had discovered was a gauge artifact or not.

To resolve this uncertainty they then considered a second choice of gauge for which the gauge-fixing term was

$$S_{gauge} = \lim_{\alpha \to \infty} \alpha \int [\nabla^a (h_{ab} - 1/4 \, g_{ab} \, h^c{}_c)]^2 \, d(Vol). \qquad (1.4)$$

In this case they also found an infra-red divergence in the propagator. They then carried out a tree-level scattering-amplitude calculation, and found an infinite result. This, they claimed, was proof that the infra-red divergence that they had found for two different gauge choices was truly a physical effect and not merely a gauge artifact.

We are going to concentrate on the second choice of gauge-fixing term (1.4) and will reach very different conclusions than those of Antoniadis et al. We are going to consider gauge-fixing terms of the following form, with $\alpha = 1/2$,

$$S_{gauge} = \alpha \int [\nabla^a (h_{ab} - \varepsilon g_{ab} \, h^c{}_c)]^2 \, d(Vol) \qquad (1.5)$$

for all values of the constant ε . We are going to prove the following three statements :

1. The graviton propagator is finite if ε does not equal one of the following "exceptional values" (1/4, 7/10, 5/6, ...)

$$\varepsilon_{exceptional} = (n^2 + 3n - 3) / (n^2 + 3n) \qquad (1.6)$$

for n = 1, 2, 3 .... If $\mathcal{E}$ has one of the exceptional values, then the propagator is infinite.

2. If $\mathcal{E}$ takes one of the exceptional values, then the propagator diverges because for that value of $\mathcal{E}$ the gauge-fixing term is not "sensitive" to a gauge transformation corresponding to a particular (finite set of) vector fields V^a.

3. The scattering amplitude is finite and independent of the value of $\mathcal{E}$.

While the gauge fixing term that we consider has $\alpha = 1/2$ and not $\alpha \to \infty$, our results apply equally well to the $\alpha = \infty$ gauge of Antoniadis et al. The reason why is because for the exceptional values of $\mathcal{E}$ the gauge-fixing-term fails to fix the gauge for _any_ value of α . In other words, when $\mathcal{E}$ takes one of the exceptional values (regardless of α) then there exist certain perturbations h_{ab} which are pure gauge $h_{ab} = \nabla_{(a}V_{b)} \neq 0$ and such that $S_{gauge}[\nabla_{(a}V_{b)}]$ vanishes. Thus our conclusion will be that Antoniadis et al. found an infra-red divergence only because they had the bad luck to choose an ineffective gauge-fixing term, and not because the graviton propagator in de Sitter space has any intrinsic physical infra-red divergence. For most choices of gauge the propagator would have been completely finite.

2.   How to find the Graviton Propagator.

The basic idea of this section is to find a form for the graviton propagator which will make it easy for us to see how it depends upon the choice of a gauge-fixing term. For this purpose it turns out to be very convenient to represent the propagator as a mode sum. Such mode sums are very familiar in the context of Lorentzian space-time calculations of (for example) the commutator and symmetric functions for a scalar field. Here we are doing something slightly less familiar - a Euclidean mode sum. The point is this : in the Hawking-Gibbons vacuum state, which is de Sitter invariant, the two-point function only depends upon the geodesic distance between the two-points. It is also an analytic function for spacelike-separated points. Therefore if we can find this function for spacelike separations, its analytic continuation to timelike separation will yield all information about the physically interesting function, which is the Lorentzian two-point function.

Thus we will look for the two-point function only for spacelike separated points. One way to do this is to carry out the calculation on a Euclidean (++++) metric 4-sphere of radius a, whose scalar curvature R has the same constant value as the curvature of the physical Lorentzian de Sitter space R = 4Λ. On this four-sphere the distance between two points is always positive, so that spacelike separation is the only possibility. It can be easily shown |18| that for such spacelike separation the two-point function on the sphere, considered just as a function of geodesic distance, is exactly the same as the Lorentzian two-point function for spatial separations. For that reason, from this moment on, we will perform all calculations on a four-sphere of radius a and volume $8\pi^2 a^4/3$. The cosmological constant is then $\Lambda = 3a^{-2}$.

To find the two-point function, we need to know the quadratic part of the gravitational action for a small metric fluctuation h_{ab}. This has been calculated in many places |19|. When we add to it the gauge-fixing term previously given (1.5), we find that the total gauged-fixed action is

$$S + S_{gauge} = (64\pi G)^{-1} \int h_{ab} W^{abcd} h_{cd} \, d(Vol). \qquad (2.1)$$

Here the second-order differential operator W^{abcd} is given by

$$W_{abcd} = ([1-2\varepsilon^2] g_{ab}g_{cd} - g_{c(a}g_{b)d}) \, \Box + (2g_{c(a}g_{b)d} + g_{ab}g_{cd}) \frac{\Lambda}{3}$$
$$+ (2\varepsilon -1)(g_{ab} \nabla_{(c} \nabla_{d)} + g_{cd} \nabla_{(a} \nabla_{b)}) \qquad (2.1)$$

where ε is the gauge-fixing parameter. Now the propagator is defined by the differential equation

$$W_{abcd} \, G^{cda'b'}(X,X') = \delta_{(a}{}^{a'} \delta_{b)}{}^{b'}, \qquad (2.3)$$

together with the boundary conditions that $G^{aba'b'}$ depend only upon the distance from X to X' (in the sense of |18|) and that it only be singular if X = X'.

This equation can be solved in many ways. One method that we have already exploited is to actually perform the path integral (1.3). This is done by choosing 10 "coordinates" in the space of all metric perturbations h_{ab}, and then integrating all of them from $-\infty$ to ∞ |20|. A simpler method will be followed here ; it involves using an ansatz which is justified by the previous method.

To use the simplest method, it is only necessary to have an orthogonal expansion of the delta function appearing on the right hand side of (2.3). This orthogonal expansion is

$$\delta_{(a}{}^{a'} \delta_{b)}{}^{b'} = \sum_{n=0}^{\infty} h_{ab}^n(X) h_n^{a'b'}(X') + \sum_{n=1}^{\infty} v_{ab}^n(X) v_n^{a'b'}(X') + \sum_{n=2}^{\infty} w_{ab}^n(X) w_n^{a'b'}(X')$$
$$+ \sum_{n=0}^{\infty} \chi_{ab}^n(X) \chi_n^{a'b'}(X') . \qquad (2.4)$$

Here the tensor fields h_n^{ab}, v_n^{ab}, w_n^{ab} and χ_n^{ab} are all eigenfunctions of $\Box$, and they form a complete set for the representation of any symmetric rank-two tensor. What this means is that any such tensor, Q_{ab}, can be represented as a sum of the form

$$Q_{ab} = \sum_0^{\infty} \alpha_n h_{ab}^n + \sum_1^{\infty} \beta_n v_{ab}^n + \sum_2^{\infty} \gamma_n w_{ab}^n + \sum_0^{\infty} \delta_n \chi_{ab}^n \qquad (2.5)$$

for a unique set of constant coefficients $\{\alpha_n, \beta_n, \gamma_n, \delta_n\}$.   If the eigenfunctions are appropriately normalized, so that

$$\delta_{nm} = \int h_{ab}^n(X) h_m^{ab}(X) \, dV = \ldots = \int \chi_{ab}^n(X) \, \chi_m^{ab}(X) \, dV \quad ,$$

$$0 = \int h_{ab}^n(X) \, v_m^{ab}(X) \, dV = \ldots = \int w_{ab}^n(X) \, \chi_m^{ab}(X) \, dV \quad , \qquad (2.6)$$

then it is easy to show from equations (2.5) and (2.6) that the delta function defined by (2.4) satisfies

$$Q^{a'b'}(X') = \int Q^{ab}(X) \, \delta^{a'}_{(a}(X,X') \delta^{b'}_{b)}(X,X') \, dV_X \qquad (2.7)$$

for any symmetric tensor Q_{ab}.

The nice thing about this method is that we don't have to explicitly construct any of the eigenfunctions $h_n^{ab}, \ldots, \chi_n^{ab}$. We will only need basic information about their eigenvalues and multiplicity which can be found in many articles [21] and which can be obtained entirely from the group representation theory of SO(5). The basic facts are as follows. The details, including the normalizations and eigenvalues of these modes, can be found in reference [20]. The ten degrees of freedom in the symmetric tensor Q_{ab} (2.5) are divided among the different modes. There are five degrees of freedom in the tensor modes h_n^{ab}, which are transverse and traceless (TT).

$$0 = h_n^{ab} \, g_{ab} = \nabla_a h_n^{ab}$$

$$\square \, h_n^{ab} = \lambda_n^{(2)} \, h_n^{ab} \qquad (2.8)$$

$$\lambda_n^{(2)} = -\frac{\Lambda}{3} (n^2 + 7n + 8)$$

There are three degrees of freedom in the modes v_n^{ab} which are the symmetrized derivatives of transverse vectors.

$$v_n^{ab} = \left[-\frac{1}{2} (\lambda_n^{(1)} + \Lambda) \right]^{-1/2} \nabla^{(a} \xi_n^{b)}$$

$$0 = \nabla_a \xi_n^a \qquad (2.9)$$

$$\square \, v_n^{ab} = (\lambda_n^{(1)} + \frac{5}{3} \Lambda) \, v_n^{ab}$$

$$\lambda_n^{(1)} = -\frac{\Lambda}{3} (n^2 + 5n + 3)$$

There is one degree of freedom in the modes w_n^{ab} which are the traceless derivatives of longitudinal vectors

$$w_n^{ab} = \left[\lambda_n^{(0)} (3/4 \, \lambda_n^{(0)} + \Lambda) \right]^{-1/2} (\nabla^a \nabla^b - \tfrac{1}{4} \, g^{ab} \, \Box) \, \phi_n$$

$$\Box \, w_n^{ab} = (\lambda_n^{(0)} + 8/3\Lambda) \, w_n^{ab}. \tag{2.10}$$

Finally there is one degree of freedom in the pure-trace scalar modes χ_n^{ab}.

$$\chi_n^{ab} = 1/2 \, g^{ab} \, \phi_n$$

$$\Box \chi_n^{ab} = \lambda_n^{(0)} \chi_n^{ab} \tag{2.11}$$

$$\lambda_n^{(0)} = -\frac{\Lambda}{3} \, n \, (n+3)$$

From a detailed examination of these modes and their normalizations (for which see |20|) it can be shown that

$$\nabla_a \nabla_b \, w_n^{ab} = \lambda_n^{(0)}(3/4\lambda_n^{(0)} + \Lambda) \, \phi_n$$

$$(g_{ab} \nabla_c \nabla_d + g_{cd} \nabla_a \nabla_b) \, \chi_n^{cd} = 2[\lambda_n^{(0)}(3/4\lambda_n^{(0)} + \Lambda)]^{\frac{1}{2}} \, w_{ab}^n + 2\lambda_n^{(0)}\chi_{ab}^n . \tag{2.12}$$

This is all that we will need to know about the mode functions.

As we said earlier, the path integral calculation of the propagator |20| justifies our use of the following ansatz for the propagator :

$$G^{aba'b'}(X,X') = \sum_0^\infty \left\{ \alpha_n \, h_n^{ab}(X) \, h_n^{a'b'}(X') \; + \right.$$
$$\beta_n \, v_n^{ab}(X) \, v_n^{a'b'}(X') \; +$$
$$\gamma_n \, w_n^{ab}(X) \, w_n^{a'b'}(X') \; + \tag{2.13}$$
$$\delta_n \, \chi_n^{ab}(X) \, \chi_n^{a'b'}(X') \; +$$
$$\left. \sigma_n \left[\chi_n^{ab}(X) w_n^{a'b'}(X') + w_n^{ab}(X) \, \chi_n^{a'b'}(X') \right] \right\} .$$

Applying the wave operator W_{abcd} to the propagator (2.13) and demanding that it yield the delta function (2.4) gives linear equations that uniquely determine the coefficients $\alpha_n, \ldots, \sigma_n$ to be

$$\alpha_n = K(-\lambda_n^{(2)} + 2/3\Lambda)^{-1} \qquad \text{for } n > 0$$

$$\beta_n = K(-\lambda_n^{(1)} - \Lambda)^{-1} \qquad \text{for } n \geqslant 1, \text{ else zero.}$$

$$\gamma_n = K(-\lambda_n^{(0)}(1-\varepsilon) + \Lambda)^{-2} \left[(2\varepsilon^2 - \varepsilon - 1/4) \lambda_n^{(0)} - \Lambda/2 \right] \text{ for } n > 2, \text{ else zero.}$$

$$\tag{2.14}$$

$$\delta_n = K(1/4\lambda_n^{(0)} + \Lambda/2)(\lambda_n^{(0)}(1-\varepsilon) + \Lambda)^{-2} \text{ for } n \geqslant 0 \tag{2.14}$$

$$\sigma_n = K(\lambda_n^{(0)}(1-\varepsilon) + \Lambda)^{-2}(\varepsilon-1/2)(\lambda_n^{(0)}(3/4\lambda_n^{(0)}+\Lambda))^{\frac{1}{2}} \text{ for } n \geqslant 2, \text{ else zero.}$$

Here the constant $K = 64\pi G$ where G is Newton's constant. With the propagator now determined by (2.13) and (2.14), we can discuss its infra-red behavior.

3. Infra-red Behavior of the Graviton Propagator.

We begin the discussion of the infra-red behavior of the graviton propagator by asserting that the propagator is finite for separated points if and only if all of the coefficients $\alpha_n, \ldots, \sigma_n$ are finite. This is indeed the case, provided that the gauge-fixing parameter ε does not have one of the values

$$\varepsilon_{exceptional} = 1 + \Lambda/\lambda_n^{(0)} = \frac{n^2+3n-3}{n(n+3)} \tag{3.1}$$

for $n = 1, 2, \ldots, \infty$. Now let us prove our assertion.

The first terms in the mode sum for the propagator, corresponding to $\alpha_n h_n h_n'$ and $\beta_n v_n v_n'$, have been evaluated by Allen and Turyn |22| and shown to be completely finite. This leaves the final three terms, which can be related to the scalar propagator, for different values of the scalar mass. Thus to understand the infra-red behavior of the graviton propagator, all we have to do is understand the scalar case.

Here the situation is very simple. For two points X and X', separated by a geodesic distance $\mu(X,X')$, the massive scalar propagator is |18,20,22|,

$$G(m^2,\mu) = \sum_{\bullet}^{\infty} \frac{\varphi_n(X)\varphi_n(X')}{-\lambda_n^{(0)} + m^2} = \frac{\Gamma(3/2 + V)\Gamma(3/2-V)}{16\pi^2 a^2} F(3/2+V,3/2-V;2;\cos^2(\mu/2a)). \tag{3.2}$$

The right hand side of this equation, and hence the mode sum, is completely finite provided that $3/2-V$ is a not a nonpositive integer. Since $V = (9/4 - a^2m^2)^{\frac{1}{2}}$, this means the propagator is finite provided that m^2 does not take one of the (negative) values

$$m^2 = -\frac{1}{a^2} n(n+3) \quad \text{for } n = 0, 1, 2, \ldots. \tag{3.3}$$

But these are exactly the values of m^2 for which the denominator $-\lambda_n^{(0)} + m^2$ in the mode sum vanishes ! Exactly the same analysis applies to the "scalar" parts of the graviton propagator. We have thus proved that provided that if ε is not given one of the "exceptional" values given above (3.1), the propagator is completely finite. What we will now do is to show why this is.

4.   How Can The Gauge-Fixing Term Fail ?

The infra-red divergence that occurs in the propagator for the exceptional values of ε can be easily understood. Imagine expressing the propagator as a path integral, or average, over all field configurations. If the gauge-fixing term was not present, then this integral would yield infinity, because it would include an infinite number of gauge-equivalent field configurations which had the same value of the action. The purpose of the gauge fixing term is to make the integral converge by giving gauge-equivalent field configurations _different_ values of the action. Thus the gauge-fixing term "fails to do its duty" if there exist a distinct pair of configurations which are physically gauge equivalent and which have the _same_ value of the gauge-fixed action. Let us now show that this is exactly what happens if ε is given one of the "exceptional" values.

We can write the gauge fixing term (1.5) in the following form, after integrating by parts.

$$S_{gauge} = -\alpha \int (h^{ab} - \varepsilon g^{ab} h^e_e) \nabla_a \nabla^c (h_{bc} - g_{bc} h^d_d) \, d(Vol). \qquad (4.1)$$

Now consider the following gauge transformation : $h^{ab} \to h^{ab} + \nabla^{(a} v^{b)}$ where $v^b = \nabla^b \varphi_n$ for the scalar mode φ_n, and $n \geqslant 1$. It is easy to verify that for $n > 0$, $\nabla^{(a} v^{b)}$ is nonzero, and

$$S_{gauge}[\nabla^{(a} v^{b)}] = 2\alpha [\lambda_n^{(0)}(1-\varepsilon)+\Lambda]^2 = 2\alpha \frac{\Lambda^2}{9} n^2(n+3)^2 [\varepsilon - \frac{n^2 + 3n - 3}{n(n+3)}]^2 . \qquad (4.2)$$

Thus, if ε takes on one of the exceptional values - say the n'th exceptional value - then the gauge-fixing term fails to be sensitive to the gauge transformation $h_{ab} \to h_{ab} + \nabla_a \nabla_b \varphi_n$ induced by the n'th scalar mode, because the r.h.s. vanishes! This is the source of the infra-red divergence that occurs for the exceptional values of ε. We will now show that this infra-red divergence, should it happen to arise because of a bad choice of ε, is a harmless gauge artifact and makes no contribution to scattering amplitudes.

5.   The Infra-red Divergence is a Gauge-Artifact.

Consider the tree level scattering process where two matter fields, which we denote ψ, interact by exchanging a graviton. Here ψ could be any kind of matter, not just a scalar field. Schematically this looks like :

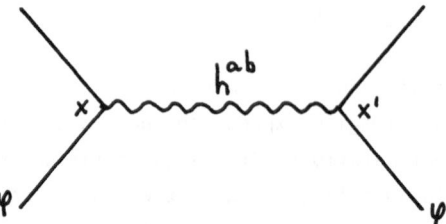

The amplitude for this process is determined by the stress tensor T^{ab} of the matter. It is

$$A = \int\int T^{ab}(X)\, G_{abc'd'}(X,X')\, T^{c'd'}(X')\, dVdV'\ ,\qquad (5.1)$$

where dV denotes the invariant four-volume element $\sqrt{g(X)}\, d^4X$ at the point X, and dV' denotes the same thing at X'. Let us assume only that $\nabla_a T^{ab} = 0$; ie that the operator T^{ab}, which is quadratic in the field ψ , is conserved. This is true even in the presence of trace anomalies, for the renormalized operator, provided that it is a matrix element between physical (on-shell) states $|23|$. We will show that this amplitude is finite regardless of the value of the gauge-fixing parameter ε, and in particular for the "exceptional" values of ε , for which $G_{abc'd'}$ contains infra-red divergences.

The amplitudes A is a sum of five terms arising from the propagator (2.13). The first two terms are independent of ε . The final three terms, upon integration by parts, can be expressed as

$$A_\gamma + A_\delta + A_\sigma\ =\ \int\int T(X)\,\rho_\varepsilon(X,X')\, T(X')\, dVdV' \qquad (5.2)$$

where $T(X) = T^a{}_a(X)$ is the trace of the stress tensor. The function $\rho_\varepsilon(X,X')$ is of the form

$$\rho_\varepsilon(X,X') = C_1 + C_2(\varepsilon - 1/4)^{-2}\,\psi_1(X)\,\psi_1(X') + C_3 \sum_{n=2}^{\infty} \frac{\psi_n(X)\psi_n(X')}{\lambda_n^{(0)} + \frac{4}{3}\Lambda}\ .$$

Here C_1, C_2 and C_3 are nonzero constants. What matter is that there appears to be a single term in the amplitude that depends upon ε . However from gauge-invariance we known that the amplitude can not depend upon ε at all ! We will now show that the second term above contributes nothing, even in the limit $\varepsilon \to 1/4$!

The reason why is simple : the mode(s) $\psi_1(X)$ obey $\nabla_a\nabla_b\psi_1 = -\frac{1}{3}\Lambda g_{ab}\psi_1$ $|24|$. Thus replacing $T^a{}_a\psi_1$ by $T^{ab}\nabla_a\nabla_b\psi_1$, and integrating by parts $|25|$ to get $(\nabla_a T^{ab})\nabla_b\psi_1$, we see that the ε-dependent term vanishes as long as the stress tensor is conserved. What this means is that even in those cases where the two-point function has an infra-red divergence, the scattering amplitude is finite. This shows that in those cases where it occurs, infra-red divergence is a harmless gauge artifact.

6.   Conclusion.

What has been shown in this talk is that the graviton propagator in de Sitter space is OK. If one makes a bad choice of gauge (-fixing term) then the propagator is
infra-red divergent. However this is not a problem. You can either make a better choice of gauge (of which there are an infinite number), for which the propagator is completely finite, or else you can go right ahead and use the infra-red divergent one.
We demonstrated that it doesn't matter. Gauge-invariance is the over-riding principle,
and it ensures that even if the propagator has an infra-red divergence, the physical
scattering amplitudes are finite.

A more detailed discussion of these points can also be found in an earlier publis-
hed paper |20|. The complete closed form for the graviton propagator with $\mathcal{E}$ = 1/2
has also been found |22|. Finally a closed form in the de Sitter -non-invariant gauge
(1.1) has been   recently obtained |26|. This form applies to any spatially-flat
Robertson-Walker model.

Acknowledgements.

I would like to thank S. Coleman, J. Iliopoulos and M. Turyn for helpful discussions.

REFERENCES

1 C.W. Misner, K.S. Thorne and J.A. Wheeler, Gravitation (Freedman, San Francisco,
   1973) p. 410.
2 I. Antoniadis, J. Iliopoulos, T.N. Tomaras, Nucl.Phys. B261 (1985) 157.
   I. Antoniadis and N.C. Tsamis, Phys.Lett. 144B (1984) 55.
   E. Baum, Phys.Lett. 133B (1983) 185.
3 G. Gibbons, S.W. Hawking and S.T.C. Siklos, The Very Early Universe, Proceedings
   of the Nuffield Workshop (Cambridge UP, 1983).
   R. Brandenburger, Rev.Mod.Phys. 57 (1985) 1.
4 B. deWit and R. Gastmans, Nucl.Phys. B128 (1985) 1.
5 N.P. Myhrvold, Phys.Lett. 132B (1983) 308.
   N.P. Myhrvold, Phys.Rev. D28 (1983) 2439.
   E. Mottola, Phys.Rev. D31 (1985) 754.
   E. Mottola, Phys.Rev. D33 (1986) 1616.
   E. Mottola, NSF-ITP 85-33 preprint UCSB.
   E. Mottola and P. Mazur, NSF-ITP 85-153 preprint UCSB.
   S. Wada and T. Azuma, Phys.Lett. 132B (1983) 313.
6 P. Anderson, University of Florida at Gainesville preprint, 1985.
7 Gary T. Horowitz, Phys.Rev. D21 (1980) 1445.
8 G.W. Gibbons and S.W. Hawking, Phys.Rev. D15 (1977) 2738.
9 B. Allen, Ann.Phys. 161 (1985) 152.
1 B. Allen, Nucl.Phys. B226 (1983) 228.
10 S.W. Hawking and G.F.R. Ellis, The Large Scale Structure of Spacetime (Cambridge
   UP, 1980).
11 O. Nachtmann, Commun.Math.Phys. 6 (1967) 1.
   N.A. Chernikov and E.A. Tagirov, Ann.Inst. Henri Poincaré IX (1968) 109.

J. Géhéniau and Ch. Schomblond, Bull.Cl.Sci., V.Ser.Acad.R.Belg. 54 (1968) 1147.
E.A. Tagirov, Ann.Phys. 76 (1973) 561.
P. Candelas and D.J. Raine, Phys.Rev. D12 (1975) 965.
Ch. Schomblond and P. Spindel, Ann.Inst. Henri Poincaré XXV (1976) 67.
T.S. Bunch and P.C.W. Davies, Proc.Roy.Soc.Lond. A360 (1978) 117.
B. Allen, Phys.Rev. D32 (1985) 3136.
B. Allen and T. Jacobson, Commun.Math.Phys. 103 (1986) 669.
B. Allen and C.A. Lütken, Commun.Math.Phys. 106 (1986) 201.

12 O. Nachtman in reference 11.
O. Nachtman, Z. Phys. 208 (1968) 113.
O. Nachtman, Sitzungsber. Oesterr.Akad.Wiss.Math.Naturwiss.Kl. 167 (1968) 363.
G.W. Gibbons and M.J. Perry, Proc.R.Soc.Lond. A358 (1978) 467.

13 I. Antoniadis, J. Iliopoulos and T.N. Tomaras, Phys.Rev.Lett. 56 (1986) 1319.

14 C. Itzykson and J.B. Zuber, Quantum Field Theory (McGraw-Hill, NY, 1980).

15 S. Coleman and E.J. Weinberg, Phys.Rev. D7 (1973) 1888.

16 This of course is the infinitesimal form of the gauge transformation.
To generate finite transformations we have to go to higher order in V.

17 The Fadeev-Popov determinant $|h^{ab}|$ does not depend upon h^{ab} at one-loop, and thus does not contribute to the tree-level propagator. We have therefore left this Jacobian out of the formula for $G_{aba'b'}$.

18 B. Allen and T. Jacobson in reference 11.

19 S.M. Christensen and M.J. Duff, Nucl.Phys. B170 (1980) 480.
N.H. Barth and S.M. Christensen, Phys.Rev. D28 (1983) 1876.
B. Allen, Phys. Rev. D34 (1986) 3670.

20 B. Allen in reference 19.

21 S.L. Adler, Phys.Rev. D6 (1972) 3445, D8 (1973) 2400.
R. Raczka, N. Limic and J. Nierderle, J.Math.Phys. 7 (1966) 1861, 7 (1966) 2026, 8 (1967) 1079.
G.W. Gibbons and M.J. Perry, Nucl.Phys. B146 (1978) 90.
S.M. Christensen, M. Duff, G.W. Gibbons, and M.J. Perry, Phys.Rev.Lett. 45 (1980) 161.
A. Higuchi, Yale Preprint YTP 85-22 (1985).
A. Chodos, E. Meyers, Ann.Phys. (NY) 156 (1984) 412.
M.A. Rubin and C.R. Ordonez, J.Math.Phys. 25 (1984) 2888, 26 (1985) 65.

22 B. Allen and M. Turyn, The graviton propagator in maximally symmetric spacer, Tufts University preprint (1986).

23 R. Wald, Phys.Rev. D17 (1978) 1477.
R. Wald, Commun.Math.Phys. 54 (1977) 1.
S.A. Fulling, M. Sweeny and R. Wald, Commun.Math.Phys. 63 (1978) 259.

24 In fact the mode that we have labeled ϕ_1 is degenerate. There are five such modes with the same eigenvalue. If the four-sphere is $X_1^2 + \ldots + X_5^2 = 1$ then the modes ϕ_1^i are proportional to the i'th coordinate X_i.

25 The boundary terms can be shown to vanish in the Lorentzian spacetime case -see reference 20.

26 B. Allen, The graviton propagator in homogeneous and isotropic spacetimes, Tufts University Preprint TUTP 86-14 (1986).
(submitted to Nucl. Phys.)

EFFECTS OF GRAVITON PRODUCTION
IN INFLATIONARY COSMOLOGY

Diego D. Harari[1]
Physics Department
Brandeis University
Waltham, MA 02254

ABSTRACT

A quantum derivation of the spectrum of gravitons created in an inflationary cosmology is discussed, and the way in which they can affect the isotropy of the cosmic microwave background is briefly reviewed.

INTRODUCTION

An inflationary cosmological model [1], in which the early universe underwent a period of exponential expansion, solves in a very attractive way many longstanding cosmological puzzles, such as the large scale isotropy and spatial flatness of our presently observed portion of the universe. Were this picture of the early universe correct, then two regions located in opposite directions in the sky that just recently entered into our Hubble sphere would have been in close causal contact in the past. So close, in fact, that quantum effects acting on such scales at early periods during the inflationary epoch, when they were small enough, may have had important consequences much later on the history of the universe. It was indeed suggested that quantum fluctuations in the energy-density during inflation might be the origin of the primordial seeds essential to explain galaxy formation [2]. At least, a Zeldovich spectrum of gaussian fluctuations naturally arises in most inflationary models, although it is not clear at present what natural inflation- driving mechanism will provide them with the adequate amplitude ($\delta\rho/\rho \approx 10^{-4}$ at horizon crossing).

[1]Supported by a Fellowship from the Consejo Nacional de Investigaciones Científicas y Técnicas, República Argentina.
Present Address: Physics Department, University of Florida, Gainesville, FL 32611.

Inflation predicts fluctuations not only in the energy-density, but also in the background metric of the space-time itself [3-6]. These metric fluctuations provide an important constraint on any model of inflation. Indeed, any stochastic background of very long wavelength gravitational waves can induce anisotropies in the cosmic microwave background, since the radiation travelling through these "ripples" in the gravitational potential can be redshifted in different ways according to the path followed. Gravitational waves with a period around 1 year can also affect in a similar way the observed "timing" of millisecond pulsars [7,8,9]. The observed bounds on these quantities can be used to place constraints on the allowed metric fluctuations produced by inflation, and hence on the parameters of the model (essentially on the value of the Hubble constant during inflation).

In the present article a quantum derivation of the metric fluctuations predicted by inflation will be discussed [10], and then the literature about their potential effect on the cosmic microwave background will be reviewed.

GRAVITON PRODUCTION

In the approach to be presented here, gravitational waves exist today because the de Sitter invariant vacuum state established during the inflationary period appears as a multiparticle state when the definition of particles relevant to the present matter-dominated universe is used. In other words, an operator representing the creation of a graviton during the radiation- or matter-dominated eras is a combination of both graviton creation and annihilation operators as defined during inflation. The coefficients of this mixing are known as Bogolyubov coefficients.

Using for convenience a conformal time variable, the background metric of the inflationary model reads

$$ds^2 = S^2(\tau)[-d\tau^2 + d\vec{x}\cdot d\vec{x}] \qquad (1)$$

with
$$S(\tau) = \begin{cases} -1/\chi\tau & \text{during inflation, while} \quad \tau < -\tau_{\ast} \\[2em] 4\tau\,\tau_m/\tau_0^2 & \text{during radiation-domination, while } \tau_{\ast} < \tau < \tau_m \quad (2) \\[2em] \tau^2/\tau_0^2 & \text{during matter-domination, while } \tau > 2\tau_m \end{cases}$$

Here χ is the value of the Hubble constant during inflation, τ_0 is the present conformal time ($\tau_0 = 3t_0$), τ_m is the conformal time at the end of the radiation

-dominated era ($\tau_m = 2^{-1/2} \, 3^{2/3} \, t_m^{1/3} \, t_0^{2/3}$) and $-\tau_*$ is the conformal time when the inflationary period finishes, $\tau_* = \tau_0/2(\chi\tau_m)^{1/2}$. The only requisite imposed on the metric is that $S(\tau)$ and its first derivative be continuous at the transitions between different regimes. The unphysical discontinuities in higher derivatives, that should be avoided in more realistic models, have no consequence on the results for long wavelengths, which are the relevants for the effects to be discussed later. Notice also that the definition of conformal time used in (2) is such that it jumps at the transitions (though S and $\dot{\text{S}}$ are continuous). These discontinuities have to be taken into account when computing the Bogolyubov coefficients.

The small metric perturbations around the Robertson-Walker background can be written in terms of graviton creation and annihilation operators (in the transverse traceless gauge) as

$$h_{ij} = \sqrt{8\pi G} \, \sum_\lambda \int \frac{d^3k}{(2\pi)^{3/2}S(\tau)\sqrt{2k}} \left[a_\lambda(\vec{k})\epsilon_{ij}(\vec{k},\lambda)e^{i\vec{k}\cdot\vec{x}} \, \xi(k\tau) + \right.$$

$$\left. + a_\lambda^\dagger(\vec{k})\epsilon_{ij}^*(\vec{k},\lambda)e^{-i\vec{k}\cdot\vec{x}} \, \xi^*(k\tau) \right] \tag{3}$$

where λ runs over the two possible polarizations and $\epsilon_{ij}(\vec{k},\lambda)$ are polarization tensors. Each independent degree of freedom can be quantized as if it were a minimally coupled scalar field [11]. The function ξ ($k\tau$) is given by

$$\xi(k\tau) = e^{-ik\tau}(1 - i/k\tau) \tag{4}$$

during inflation and matter-domination, and by

$$\xi(k\tau) = e^{-ik\tau} \tag{5}$$

during the radiation-dominated era.

The graviton creation and annihilation operators at differents eras are related through

$$a_\lambda^{rad}(\vec{k}) = \alpha_1(k)a_\lambda^{inf}(\vec{k}) + \beta_1^*(k)\left[a_\lambda^{inf}(-\vec{k})\right]^\dagger$$

$$a_\lambda^{mat}(\vec{k}) = \alpha_2(k)a_\lambda^{inf}(\vec{k}) + \beta_2^*(k)\left[a_\lambda^{inf}(-\vec{k})\right]^\dagger \tag{6}$$

These Bogolyubov coefficients can be evaluated matching both h_{ij} and its first derivative at both transitions. The result is, for modes well outside the horizon at τ_* or τ_m respectively ($k\tau_* \ll 1$ or $k\tau_m \ll 1$)

$$\alpha_1 \approx -\beta_1 \approx \frac{-2\chi\tau_m}{(k\tau_0)^2}$$

$$\alpha_2 \approx \beta_2 \approx \frac{-3i\chi}{2k^3\tau_0^2} \tag{7}$$

The quantum state of the system will be characterized in terms of the number of gravitons as defined during inflation. In other words, eigenstates of the operator $(a_\lambda^{inf})^\dagger a_\lambda^{inf}$ will be considered. This way of characterizing the quantum state $|\psi\rangle$ may seem arbitrary, but recall that the wavelengths of relevance for the cosmic microwave background anisotropy are of the order of the size of the universe today, that is $k\tau_0 \approx 1$. For these waves, early during the inflationary period $|k\tau| \gg 1$, which means that they were originally well inside the horizon, in which case the modes given by (4) are just the ordinary Minkowski definition of positive frequency modes.

In the quantum state $|\psi\rangle$, the fluctuations in h_{ij} at time τ and wavenumber $\vec{k}$ can be characterized by

$$\Delta h^2(\vec{k}) \equiv \frac{k^3}{(2\pi)^3} \cdot \frac{1}{2} \int d^3\vec{x}\, e^{i\vec{k}\cdot\vec{x}} \langle\psi|h_{ij}(\vec{x},\tau)h_{ij}(\vec{0},\tau)|\psi\rangle \tag{8}$$

with a sum on i and j understood. Wavelengths much larger than the present observable universe have to be cut-off when actually computing this expression. Choosing the quantum state $|\psi\rangle$ to be the vacuum state (which is de Sitter invariant) and using the relevant Bogolyubov coefficients given in (7), the fluctuations for long wavelengths ($k\tau \ll 1$) turn out to be

$$\Delta h^2(k) \approx \frac{G\chi^2}{2\pi^2}\left[\frac{\sin k\tau}{k\tau}\right]^2 \text{ during radiation-domination}$$

$$\Delta h^2(k) \approx \frac{9G\chi^2}{2\pi^2}\left[\frac{J_1(k\tau)}{k\tau}\right]^2 \text{ during matter-domination} \tag{9}$$

Notice that in both cases the amplitude of the fluctuations tend to a constant value, $G\chi^2/2\pi^2$, for wavelengths outside the horizon, in agreement with alternative derivations [3-6].

If instead of choosing the state $|\psi\rangle$ as the vacuum, a state with $N_\lambda(\vec{k})$ particles is considered, the result (9) would appear multiplied by a factor $[1+N_\lambda(\vec{k})+N_\lambda(-\vec{k})]$. Since at the start of inflation the wavelengths under consideration were well inside the horizon, it seems more than plausible to assume $N_\lambda(k)\ll1$. Otherwise the energy density in what at that time were high frequency modes would be enormous.

EFFECTS UPON THE COSMIC MICROWAVE BACKGROUND

Small perturbations of the metric around a Robertson-Walker background can induce an anisotropy in the observed temperature of the cosmic microwave background. The effect is described by the Sachs-Wolfe formula [12, 13]

$$\frac{\delta T_0}{T_0} = - \int_0^{\tau_0-\tau_E} dz \; \dot{h}_{ij}(z\vec{e},\tau_0-z)e^i e^j \qquad (10)$$

where τ_0 and τ_E are the conformal time for the emission and reception of the radiation respectively, T_0 is the temperature observed at present times and $\vec{e}$ is a unit vector pointing in the direction of the observation.

Knowing from the previous section how to evaluate expectation values of the field h_{ij}, the fluctuations in the temperature of the cosmic microwave background induced by the quantum fluctuations of the metric predicted by inflation can be computed. The temperature anisotropy can be expanded in a multipole decomposition as

$$\frac{\delta T_0(\theta,\phi)}{T_0} = \sum_{\ell,m} a_{\ell m} Y_{\ell m}(\theta,\phi) \qquad (11)$$

It is more convenient to work with the rotationally invariant quantities a_ℓ defined by

$$a_\ell^2 \equiv \sum_{m=-\ell}^{\ell} |a_{\ell m}|^2 \qquad (12)$$

Each $a_{\ell m}$ can be treated as a gaussian random variable [3,14,15] (since h_{ij} is a free field) with r.m.s. deviation from the zero mean given in terms of the vacuum expectation value of $a_{\ell m}^2$. The a_ℓ will then also be gaussian variables, with standard deviation determined by $\langle 0|a_\ell{}^2|0\rangle$, which can be computed. There will be no dipole contribution to the anisotropy coming from gravitational waves. The quadrupole term gives [3-6]

$$\langle 0|a_2{}^2|0\rangle \approx G\chi^2 \tag{13}$$

(the integration in the Sachs-Wolfe formula is done for wavelengths that entered the horizon after recombination). Being a_2 a gaussian random variable one can expect that $a_2{}^2 > 0.3\langle 0|a_2{}^2|0\rangle$ with a 90% confidence level. That is the prediction. The observational bound is [16] $a_2{}^2 < 5.2 \times 10^{-8}$. Consequently, with a 90% confidence level

$$(G\chi^2)^{1/2} = \frac{\chi}{M_{p\ell}} \leq 10^{-4} \tag{14}$$

Equivalently, this bound can be transformed into a bound on the maximum reheating temperature of the universe when inflation ends. Assuming a perfect reheating

$$T_{RH}(Max) \approx 10^{17} \text{ GeV} \tag{15}$$

Predictions for higher multipole moments can also be made [3,6]. They are relevant because an eventual measurement or bound on higher multipole anisotropies can also test the spectrum predicted by inflation [13,14].

GENERALIZED INFLATION

It has been pointed out [3] that the homogeneity and spatial flatness of the present universe can be explained by a long enough early period of "accelerated expansion", during which the physical distance between two points of fixed coordinate distance grew faster than the horizon size. Such property of the conventional exponential inflation is shared with many other possible scale factor evolutions [3,17,18]. For instance, if $R(t) \approx t^p$ with p>1, the requirement $\dot{R} > 0$ (that defines these generalized inflationary models) is met. The effects of metric [3] and energy-density fluctuations [18] in these and other generalized models have been studied. For completeness, here it is shown briefly how to apply the method of the previous sections to evaluate the metric fluctuations in these generalized models and the corresponding bounds on the reheating temperature.

The scale factor during the generalized inflationary period reads $R(t)=h^{p-1}$ t^p. Assuming a transition to a radiation dominated universe, the scale factor and its first derivative matched continuously, then in the conformal time variable:

$$S(\tau) = \begin{cases} \dfrac{1}{h} \, [(1-p)\tau]^{p/1-p} & \text{while } \tau < -\dfrac{p}{p-1}\, \tau_{\ast} \\[4mm] 4\tau\, \tau_m/\tau_0^2 & \text{while } \tau > \tau_{\ast} \end{cases} \tag{16}$$

where

$$\tau_{\ast} = (\frac{\tau_0^2}{4\tau_m h})^{1-p/1-2p} \cdot p^{p/1-2p} \qquad$$

The normal modes of the graviton field decomposition (3) now are, during the generalized inflationary period,

$$\xi(k\tau) = (\frac{\pi k\tau}{2})^{1/2} \, H_{\nu}^{(2)}(k\tau); \quad \nu \equiv \frac{3p-1}{2(p-1)} \tag{17}$$

They behave like ordinary positive frequency modes in the limit $|k\tau| \gg 1$. The Bogolyubov coefficients that relate the graviton creation and annihilation operators during a radiation-dominated period to those of the inflationary regime behave, when $k\tau_{\ast} \ll 1$, as

$$\alpha(k) \approx -\beta(k) \approx \frac{e^{-i\pi\nu}\Gamma(\nu)}{(2\pi)^{1/2}} \cdot 2^{(\nu-1)} \cdot (\nu - \frac{1}{2})^{(1/2-\nu)} \cdot (k\tau_{\ast})^{-(\nu + 1/2)} \tag{18}$$

The expression (8) can now be used to evaluate the vacuum expectation value of the metric fluctuations for long wavelengths during the radiation-dominated period with the result

$$\Delta h^2(k) \approx \frac{Gh^2\Gamma^2(\nu)\, 2^{2\nu-2}}{\pi^3(p-1)^{1-2\nu}k^{2\nu-3}} \, (\frac{\sin k\tau}{k\tau})^2 \tag{19}$$

or, equivalently, if we define $H_{HC}(k)$ as the value of the Hubble constant when the wavelength of coordinate wavenumber k left the horizon during inflation $(k/S(\tau_{HC})H_{HC} = 1)$

$$\Delta h^2(k) = \frac{GH_{HC}^2(k)}{2\pi^2} \cdot \frac{\Gamma^2(\nu)}{\pi} \left[\frac{2(p-1)}{p}\right]^{2\nu-1} (\frac{\sin k\tau}{k\tau})^2 \tag{20}$$

The spectrum is strictly scale-invariant only for the exponential inflation ($p \to \infty$). Indeed, when $\upsilon = 3/2$ the expression (19) has no explicit k-dependence: the amplitude of the fluctuations was the same for any scale at the time when that scale left the horizon. The expression (20) shows more transparently that the scale-invariance of the spectrum in the exponential inflationary cosmology is related to the constancy of the Hubble coefficient. A similar analysis can be done for the transition from the radiation-dominated to a matter-dominated regime. The amplitude of the fluctuations for wavelengths that enter the horizon during matter-domination is the same as in (20).

The Sachs-Wolfe effect can now be computed in these generalized inflationary models, and the requirement for the fluctuations not to induce a quadrupole anisotropy on the cosmic microwave background beyond observational bounds imposed. Then a bound on H_{HC}, which can be transformed into a bound on the maximum reheating temperature after inflation, is obtained. It depends, of course, on p. The upper bound on T_{RH} is smaller as $p \to 1$. For example, $T_{RH} < 10^8$ GeV if $p = 2$. See the references [3,18] for details.

CONCLUSIONS

The bound on the value of the Hubble constant during inflation, $\chi < 10^{15}$ GeV, is an important constraint on candidate models for inflation. For instance, one should not expect inflation to be a quantum gravity process, taking place at scales comparable to the Planck scale. In the case of the generalized inflationary evolutions the bound on the reheating temperature can be used to restrict possible models, if the reheating temperature is to be large enough to explain the baryon number of the universe through baryon number non conservation processes [3].

A bound can also be placed on the amplitude of an stochastic background of gravitational waves of period around 1 year using millisecond pulsar timing measurements [7-9]. The gravitational waves should not induce a noise in the signal larger than the observed r.m.s. of the residual of the arrival time of the signal, after fitting the data for period, period derivative, position and proper motion. The bound on the squared amplitude of metric fluctuations in inflationary cosmologies derived in this way is weaker than the one coming from the isotropy of the cosmic microwave background by at least four orders of magnitude. However, with an improvement in the earth-based timekeeping technology the millisecond pulsar bound may in principle improve as the fourth power of the duration of the observations [8,9].

ACKNOWLEDGEMENTS

My work on this area was done in collaboration with Larry Abbott, to whom I am grateful for introducing me to the subject. It is a pleasure to acknowledge the hospitality extended to me while I was visiting the Physics Department of Brandeis University.

REFERENCES

1. A. Guth, Phys. Rev. D23 (1981) 347.
2. A. Guth and S. Y. Pi, Phys. Rev. Lett. 115B (1982) 189.
    S. Hawking, Phys. Lett. 117B (1982) 295.
    A. Starobinskii, Phys. Lett. 117B (1982) 175.
    J. Bardeen, P. Steinhardt and M. Turner, Phys. Rev. D28 (1983) 679.
3. L. Abbott and M. Wise, Nucl. Phys. B244 (1984) 541.
4. A. Starobinskii, JETP Lett. 30 (1979) 682 and Sov. Astron. Lett. 9 (1983) 302..
5. V. Rubakov, M. Sazhin and A. Veryaskin, Phys. Lett. 115B (1982) 189.
6. R. Fabbri and M. Pollock, Phys. Lett. 125B (1983) 445.
7. S. Detweiler, Astrophys. J. 234 (1979) 1100.
    B. Mashhoon, Mon. Not. R. Astr. Soc. 199 (1982) 659.
    B. Bertotti, B. J. Carr and M. Rees, Mon. Not. R. Astr. Soc. 203 (1983) 945.
8. L. Krauss, Nature 313 (1985) 32.
9. M. Davis, J. Taylor, J. Weisberg and D. Backer, Nature 315 (1985) 547.
10. L. Abbott and D. Harari, Nucl. Phys. B264 (1986) 487.
11. L. Ford and L. Parker, Phys. Rev. D16 (1977) 1601.
    B. L. Hu and L. Parker, Phys. Lett 63A (1977) 217.
12. R. Sachs and A. Wolfe, Astrophys. J. 147 (1967) 73.
13. L. Abbott and R. Schaeffer, preprint LHEA 86-003 (1986).
    R. Fabbri, F. Lucchin and S. Matarrese, Preprint Univ. of Padua, DFPD-11/86 (1986).
14. L. Abbott and M. Wise, Phys. Lett. 135B (1984) 279.
15. L. Abbott and M. Wise, Astrophys. J. 282 (1984) 647.
16. P. Lubin, G. Epstein and G. Smoot, Phys. Rev. Lett. 50 (1983) 616.
    D. Fixen, E. Cheng and D. Wilkinson, PHys. Rev. Lett. 50 (1983) 620.
17. F. Lucchin and S. Matarrese, Phys. Rev. D32 (1985) 1316.
18. F. Lucchin and S. Matarrese, Phys. Lett. 164B (1985) 282.
    R. Fabbri, F. Lucchin and S. Matarrese, Phys.Lett. 166B (1986) 49.

MULTI-DIMENSIONAL INTEGRABLE SYSTEMS

R.S.Ward

Department of Mathematics
Durham University
Durham, DH1 3LE, U.K.

1. Introduction

This lecture is concerned with integrable systems of differential
equations. I am not using the word "integrable" in the sense of
Frobenius' theorem: it does not refer to the existence of solutions to
the equations. Rather, it refers to a special property which certain
equations have. In the case of classical mechanics, for example, it
implies that one can transform to action-angle variables. In addition
to classical mechanics (ordinary differential equations), I shall also
be concerned with partial differential equations; examples of these
which I include in the class of integrable equations are soliton equations such as KdV and sine-Gordon, and integrable field-theory equations
such as self-dual Yang-Mills and 2-dimensional σ-models. I shall not
say anything about techniques for solving these systems; my interest
is in trying to understand what integrability _is_, in some fairly general
sense.

Integrable equations are very special: roughly speaking, almost
every equations is _not_ integrable. Nevertheless, integrable systems
turn up over and over again in the study of non-linear phenomena; and
in addition they are associated with some very beautiful mathematics.

In section 2 below I shall discuss how one might try to define what
integrability is. Section 3 describes some classes of integrable
systems which arise as reductions of the self-dual Yang-Mills equations.
I shall then consider generalizations of this scheme, in particular to
higher dimensions.

It is worth remarking that the notion of integrability also applies
to quantum field theories, and an interesting question is how this is
related to classical integrability. But in what follows, I shall
restrict to classical systems.

2. What is Integrability?

Let us first recall the situation with regard to ordinary differential equations. For Hamiltonian systems with n degrees of freedom, the classic "Liouville" definition of integrability is in terms of the existence of sufficiently many constants of motion. Namely, there should exist on phase space n independent functions K_i which Poisson-commute with the Hamiltonian and with one another. It is important to emphasize that such functions always exist <u>locally</u> on phase space (except possibly around critical points of the Hamiltonian). So integrability is the statement that they should exist <u>globally</u>, in some appropriate sense.

One can tie things down with a stronger definition such as "algebraic complete integrability" (ACI). This involves complexifying the system, so that the phase-space variables become complex-analytic functions of complex time. The condition of ACI is that the flow is linear on abelian varieties (complex algebraic tori) defined by K_i = constant. (See van Moerbeke 1985 for a review.) This implies that the phase-space variables have the "Painlevé property", i.e. that they are meromorphic functions of complex time. And conversely, if a system has the Painlevé property, then it seems to stand a good chance of being ACI.

One problem with ACI, and with the Painlevé test, is that they are somewhat coordinate-dependent, and therefore rather restrictive. For example, a system with only one degree of freedom, with Hamiltonian $H(p,q) = \frac{1}{2}p^2 + V(q)$, may fail to satisfy ACI or the Painlevé property (if the potential function V is not sufficiently well-behaved). But such a system <u>is</u> integrable in the classical sense.

It has been suggested that one should consider a wider class of dynamical systems which are in general non-integrable (in fact ergodic), but nevertheless "solvable". See, for exmaple, Eckhardt et al (1984) and Chudnovsky (1984), who discuss systems associated with "rational billiards in the plane", or geodesics on surfaces of negative curvature, as examples of such solvable systems. I shall not consider such systems here.

The subject of ordinary differential equations will come up again later, but for the remainder of this section I shall study the question of what is meant by integrability for a system of <u>partial</u> differential equations.

(a) By analogy with the Liouville definition, one may specify that the

system admit an infinite number of conserved currents. Certainly this
idea is relevant to soliton equations such as KdV, where one gets an
infinite number of commuting flows on an infinite-dimensional space, so
there is a close analogy with the finite-dimensional Liouville picture.
But there are a number of objections to using this as a definition of
integrability.   First, it does not seem to apply to systems such as
the self-dual Yang-Mills or self-dual Einstein equations, at least not
in any natural way.   Secondly, and more seriously, consider the "direct
sum" of a non-integrable system, and an integrable system with infinitely
many conserved densities.   The combined system has infinitely many conserved densities (which happen to depend only on the first summand), but
is not integrable (because of the second summand).   The same argument
rules out definitions involving infinitely many symmetries, Bäcklund
transformations, and the like.

It is interesting to note that any conformally-invariant 2-dimen-
sional system has an infinite number of conserved currents (Goldschmidt
& Witten 1980), which are important in the corresponding quantum field
theory.   But such systems are not in general integrable.   For example,
consider the "harmonic map" or "σ-model" problem for functions ϕ from
$\mathbb{R}^2$ to M, defined by the Lagrangian

$$L = \delta^{\mu\nu} (\partial_\mu \phi^i)(\partial_\nu \phi^j) g_{ij}(\phi),$$

where g_{ij} is the metric on M.   This is conformally invariant, but may
not be integrable.   (For example, the solutions which depend on only
one of the Cartesian coordinates of $\mathbb{R}^2$, correspond to geodesics on M.
And geodesic flow on a Riemannian manifold is seldom integrable.)

(b) One may require that the system has the Painlevé property, which
(roughly speaking) would say that its solutions should be meromorphic
functions of the complexified independent variables (see, for example,
Ward 1984).   This can be useful as a test for integrability, and for
revealing properties of integrable systems.   But it is not much good as
a definition of integrability, since for one thing it is not preserved
under a transformation of the dependent variables.

(c) Integrability seems to be a property which can be said to hold
locally in space-time.   For example, the sine-Gordon equation is just
as integrable on a flat 2-dimensional torus as it is on flat space-time
$\mathbb{R}^2$.   In particular, boundary conditions are irrelevant as far as integrability is concerned (although obviously they matter for the existence
of particular kinds of solutions such as solitons).   Thus one should
not define integrability in terms of ideas such as the existence of

multi-soliton solutions, or the inverse scattering transform, which
depend on appropriate global conditions. (Also, these ideas do not
seem to apply to systems such as the self-dual Yang-Mills equations.)

(d) Many integrable systems are closely associated with Lie algebras,
and in particular with infinite-dimensional (Kac-Moody) Lie algebras.
Algebraic structure is also important in integrable quantum field
theories and statistical-mechanics systems. But it is not clear to me
how fundamental such algebraic background is, and whether it only
applies to a specific class of integrable systems with specific boundary
conditions.

(e) One can study the reductions of a given system of partial differential equations to ordinary differential equations. For example, the
Yang-Mills equations and Einstein's equations can be reduced by (essentially) saying that the fields depend only on time (and not on the space
variables). And the resulting "mechanical" systems can exhibit chaotic
behaviour ("chaotic cosmology" in the Einstein case; see e.g. Savvidy
1984 for the Yang-Mills case). From this one should certainly conclude
that Einstein's and the Yang-Mills equations are not integrable. But
as a test (or definition) of integrability, this method may not be very
reliable, since the original system may not possess "enough" reductions.

(f) Finally, one could require that the system of equations be the
consistency condition for an overdetermined system of linear equations.
However, this linear system has to have some special property. For
example, Dubois-Violette (1982) showed that both the Einstein and the
Yang-Mills equations can be written as consistency conditions for a
linear system, but as noted above, these equations are not integrable.
So one has to restrict to "allowable" linear systems which do give rise
to integrable equations. I do not know what the most general allowable
linear systems are; in what follows, I shall merely give some examples
of allowable systems. But it seems that the best way to define "integrability" in general, would be to specify what is meant by "allowable".

3. Reductions of the Self-Duality Equations

     A basic example of an integrable system is that of the self-dual
Yang-Mills equations in four dimensions. Its solutions can be described in terms of holomorphic vector bundles, or, equivalently, in

terms of the "Riemann-Hilbert problem". But I shall not be concerned with the various solution techniques here, merely with listing some of the systems that can be obtained from the self-duality equations by reduction.

Let A_μ denote a gauge potential on complexified Minkowski space-time C^4, taking values in some complex Lie algebra $\underline{g}$. Its curvature is

$$F_{\mu\nu} = \partial_\mu A_\nu - \partial_\nu A_\mu + [A_\mu, A_\nu]$$

and the self-duality equations are

$$\tfrac{1}{2} \epsilon_{\mu\nu\alpha\beta} F^{\alpha\beta} = F_{\mu\nu},$$

which are a system of semi-linear first-order partial differential equations for A_μ. They are the consistency conditions for a pair of linear equations which may be written

$$(D_y - \zeta D_v)\psi = 0,$$

$$(D_u + \zeta D_z)\psi = 0.$$

$$(1)$$

Here $D_\mu = \partial_\mu + A_\mu$ denotes the covariant derivative corresponding to A_μ, ζ is a complex parameter (sometimes called the "spectral paramater"), and u, v, y, z are null coordinates on space-time (the metric is $ds^2 = dydz + dudv$). In terms of these coordinates, the self-duality equations are

$$F_{yu} = F_{zv} = 0, \quad F_{yz} + F_{uv} = 0. \qquad (2)$$

These follow from the requirement that (1) be consistent for all values of the parameter ζ. To reduce this system involves two processes.

(A). First, one can reduce the number of independent variables by "factoring out" by a subgroup of the Poincaré group. (The system (1) is in fact Poincaré-invariant, although I have not written it here in Poincaré-invariant form.) See Forgacs & Manton (1980) for a general discussion of what this "dimensional reduction" entails.

(B). Secondly, one can reduce the number of dependent variables by imposing algebraic conditions on the A_μ in a consistent way. (I don't know whether one should also be allowed to impose differential constraints on the A_μ, since this means in effect adding extra differential equations to the system.)

A complete analysis of all reductions of the system (1) looks like a large problem, because there are so many possibilities for both (A) and (B). I shall merely exhibit some examples.

EXAMPLE 1.   First, reduce as in (A) by factoring out the two trans-lations ∂_u and ∂_v.   In other words, impose the condition that A_μ be a function of y and z only.   To reduce as in (B), we shall use some algebra; for more details of the algebraic background, see Olive & Turok (1983).   (Their treatment is analogous to, although somewhat different from, the one presented here.)

Suppose that the gauge algebra $\underline{g}$ is a simple Lie algebra of rank n (think of $s\ell(n-1)$). Let $\{H_a, E_a, E_{-a}\}_{a=1,2,\ldots,n}$ be a Chevalley basis of $\underline{g}$, satisfying the usual commutator relations

$$[H_a,H_b] = 0, \quad [E_a,E_{-b}] = \delta_{ab}H_b,$$

$$[H_a,E_b] = K_{ba}E_b, \quad [H_a,E_{-b}] = -K_{ba}E_{-b}. \tag{3}$$

Here K_{ba} is the Cartan matrix. We can reduce the number of dependent variables A_μ by requiring that they have the form

$$A_y = \sum_a f_a H_a, \quad A_z = \sum_a g_a H_a,$$

$$A_u = \sum_a e_a E_a, \quad A_v = \sum_a e_a E_{-a}, \tag{4}$$

where f_a, g_a, e_a are functions of y and z.   Then a simple calculation shows that the self-duality equations (2) reduce to the "Toda molecule" equations

$$\partial_y \partial_z \phi_a = -\sum_b K_{ab} \exp \phi_b \tag{5}$$

where $\phi_a = 2 \log e_a$.   The simplest case is n = 1, algebra $s\ell(2)$, where $K_{11} = 2$ and one gets the Liouville equation $\partial_y \partial_z \phi = -2e^\phi$.

For a second example, let us use the extended Cartan matrix $\bar{K}_{ab}$, where the indices a and b now run from 0 to n, labelling an extended system of simple roots (0 corresponds to minus the highest root).   The commutation relations (3) remain valid, with K_{ab} replaced by $\bar{K}_{ab}$; and we use the same reduction (4), writing $e_a = \exp(\tfrac{1}{2}\phi_a)$.   I shall do this case in more detail.   First,

$$F_{yu} = \sum_{a,b} \tfrac{1}{2}e_a(\partial_y \phi_a + 2f_b \bar{K}_{ab}) E_a$$

and the E_a are linearly independent, so $F_{yu} = 0$ gives

$$\partial_y \phi_a = -2 \sum_b \bar{K}_{ab} f_b. \tag{6}$$

Similarly, $F_{zv} = 0$ gives

$$\partial_y \phi_a = 2 \sum_b \bar{K}_{ab} g_b. \tag{7}$$

And from $F_{yz} + F_{uv} = 0$ we get

$$\sum_a (\exp \phi_a - \partial_z f_a + \partial_y g_a) H_a = 0. \tag{8}$$

Because the H_a are not linearly independent, we cannot deduce that the expression in parentheses in (8) vanishes, but we can deduce

$$\sum_a \bar{K}_{ba}(\exp \phi_a - \partial_z f_a + \partial_y g_a) = 0. \tag{9}$$

Now from (6), (7), (9) we get

$$\partial_y \partial_z \phi_a = - \sum_b \bar{K}_{ab}(\partial_z f_b - \partial_y g_b)$$

$$= - \sum_b \bar{K}_{ab} \exp \phi_b \tag{10}$$

which are the so-called "Toda lattice" equations. The simplest case is when $n = 1$, algebra $s\ell(2)$, with

$$\bar{K}_{ab} = \begin{bmatrix} 2 & -2 \\ -2 & 2 \end{bmatrix},$$

so that (10) becomes (essentially) the sinh-Gordon equation $\partial_y \partial_z \phi = -4 \sinh \phi$.

A further generalization gives the "generalized Toda lattice equations", which include, for example, the Bullough-Dodd equation $\partial_y \partial_z \phi = -2e^\phi + e^{-2\phi}$. Cf. Olive & Turok (1983).

EXAMPLE 2. As before, suppose the A_μ depend only on y and z. But now, impose the constraint $A_u = A_z$, $A_v = -A_y$. Then the self-duality equations (2) reduce to

$$\partial_y A_z - \partial_z A_y + 2[A_y, A_z] = 0, \tag{11a}$$

$$\partial_y A_z + \partial_z A_y = 0. \tag{11b}$$

These are the 2-dimensional chiral field equations; they can be written in a more usual form as follows. Eqn. (11a) says that $2A_\mu$ has zero curvature, and therefore must have the form

$$A_\mu = \tfrac{1}{2} g^{-1} \partial_\mu g,$$

where g takes values in the gauge group $\underline{G}$. (The index μ is now 2-dimensional: we are in 2-dimensional space-time with metric dydz.) Eqn. (11b) then gives

$$\partial^\mu (g^{-1} \partial_\mu g) = 0,$$

which is the usual form of the chiral field equations. Imposing further algebraic constraints on the A_μ (or on g) gives examples such as CP^n-models and non-linear σ-models. Such reductions correspond to symmetric spaces, and have been completely classified.

EXAMPLE 3. Instead of factoring out translations, one could, for example, factor out rotations. In general this will lead to non-autonomous equations, i.e. equations in which the independent variables appear explicitly. An example is the "Ernst equation" of general relativity, which can be written in the form

$$\partial_z(J^{-1}\partial_z J) + \rho^{-1}\partial_\rho(\rho J^{-1}\partial_\rho J) = 0, \qquad (12)$$

with $J(\rho,z)$ being a 2×2 matrix. The solutions of (12) correspond to stationary axisymmetric solutions of Einstein's equations. One can obtain (12) by a reduction analogous to that of the previous example, except that one factors out a rotation and a translation rather than two translations. Some more details may be found in Witten (1979) and Ward (1983).

EXAMPLE 4. Finally, some one-dimensional reductions, i.e. ordinary differential equations. One reduction which has been of interest is achieved by saying that A_μ should depend only on $y + z$: this gives the Nahm equations (see, for example, Ward 1985 and the references cited therein). Let me rather give an example of what happens if we reduce by $A_\mu = A_\mu(y)$. Then the compatibility condition for (1) reduces to

$$\frac{d}{dy}(A_z\zeta + A_u) = [A_z\zeta + A_u, -A_v\zeta + A_y]. \qquad (13)$$

This is a "Lax equation with parameter", about which there is an extensive mathematical theory (Adler & van Moerbeke 1980). It is this kind of structure which really seems to lie at the heart of completely integrable mechanical systems.

Clearly (13), or alternatively (2) with $A_\mu = A_\mu(y)$, is equivalent to

$$[A_z,A_v] = 0, \qquad (14a)$$

$$A'_z = [A_z,A_y] + [A_v,A_u], \qquad (14b)$$

$$A'_u = [A_u,A_y], \qquad (14c)$$

where the prime denotes d/dy. Let us impose algebraic constraints on the A_μ by saying that A_u and A_y are skew-symmetric $n \times n$ matrices, while A_v and A_z are diagonal with

$$A_z = \text{diag}(a_1,\ldots,a_n),$$

$$A_v = \text{diag}(b_1,\ldots,b_n),$$

the a_i and b_i being constants. Then (14a) is satisfied, (14b) determines A_y in terms of A_u, and (14c) then becomes

$$M'_{ij} = \sum_k (\lambda_{ik} - \lambda_{kj})M_{ik}M_{kj}, \qquad (15)$$

where

$$\lambda_{ij} = (b_i - b_j)/(a_i - a_j) \qquad \text{for } i \neq j$$

and where M_{ij} denotes the ij-th element of the matrix A_u. Equation (15) is the Euler-Arnold-Manakov equation for an n-dimensional spinning rigid body, or alternatively the equation of geodesic flow on SO(n), with λ_{ij} defining a left-invariant diagonal metric on SO(n).

4. Generalizations of the Self-Duality Equations

One can generalize the linear system (1) in a number of different ways. Observe that (1) consists of two equations, each linear in the single spectral parameter ζ. So one could:

(i)   increase the number of equations; or

(ii) allow each equation to be a polynomial in the spectral parameter(s), of degree greater than 1; or

(iii) increase the number of spectral parameters;

or any combination of these. There is a discussion of these generalizations in Ward (1984b), and I shall merely mention here some of the features that arise.

First of all, there is an increase in the number of "space-time" dimensions. For example, if one replaces (1) by a pair of equations quadratic in ζ, namely

$$(D_y - \zeta D_v + \zeta^2 D_x)\psi = 0$$

$$(16)$$

$$(D_u + \zeta D_z + \zeta^2 D_w)\psi = 0,$$

then one is dealing with a gauge field in 6-dimensional space, with coordinates (u, v, w, x, y, z). The consistency conditions for (16) amount to a set of linear relations on the curvature $F_{\mu\nu}$, generalizing the self-duality equations (2).

Not much is known about the reductions of (16) and of its consistency conditions. However, there is a 2-dimensional reduction which is well-known, namely the non-linear Schrödinger (NLS) equation. The linear system for NLS, and also for the generalized NLS equations of Fordy & Kulish (1983), have the form (16). Similarly, the linear system for the KdV equation and its generalized versions (Athorne & Fordy 1986) is a pair of equations which are cubic in ζ. So from this point of view, the KdV equation is a reduction of a set of gauge-field equations in eight dimensions (although for practical purposes this may not be a very useful way of dealing with the KdV equation!).

Possibility (i), namely increasing the number of linear equations, also increases the number of space-time dimensions. Now, however, the

consistency conditions become (not surprisingly) overdetermined. There
is also a 2-dimensional reduction of this (or rather, of (i) combined
with (ii)) which is well-known, namely the "KdV hierarchy". This is
an overdetermined system which involves commuting flows with respect to
"many times" (see, for example, Date et al 1983).

   Possibility (iii), namely increasing the number of spectral parameters, also leads to an overdetermined system in higher dimensions.
In this case, however, it does not seem that there are any interesting
reductions to low dimensions.

   Clearly, all the 2-dimensional soliton equations which fit into
the AKNS or Zakharov-Shabat framework, also fit into the general scheme
described here. One interesting question is whether one could deal
with the "complete" KdV hierarchy, which involves infinitely many
commuting flows on an infinite-dimensional space (Date et al 1983,
Wilson 1985). It would have to involve an infinite-dimensional analogue
of the structure discussed here.

5. Other Types of Linear System

   One integrable system which does not fit into the above scheme is
that of the self-dual Einstein equations. It is, however, rather
similar in nature. The linear system can be written in the particularly
elegant form

$$\nabla_{AA'} \, \pi_{B'} = 0, \qquad (17)$$

where $\nabla_{AA'}$ is the covariant space-time derivative, and $\pi_{B'}$ is a 2-com-
ponent spinor. The consistency condition for (17) is indeed the self-
dual Einstein equation, and this description leads to Penrose's (1976)
"twistor" construction of its solutions. The spectral parameter in
(17) is the ratio of the two components of $\pi_{B'}$.

   All the consistency conditions discussed up to now have been "the
vanishing curvature of a connection". But there are some equations
that should be regarded as integrable, but that have not (to my knowledge) been represented in this way. These are exemplified by the KP
equation (Date et al 1983), the linear system of which can be written
in terms of higher-order differential operators or pseudo-differential
operators, but not (apparently) in terms of first-order differential
operators of "connection" type. This may be related to the comment
at the end of the previous section, about the need to go to an infinite-
dimensional space. But in any event, the scheme described by Date et
al and Wilson is so elegant that one may not want to replace it with
something else.

6. <u>Conclusions</u>

I began this lecture by posing the question of what the best way is to define integrability. The most promising definition seems to be one based on an associated overdetermined linear system which is "allowable". The linear systems described in sections 3 and 4 are certainly allowable, and most known integrable equations can be obtained in this way. But there are exceptions, such as the KP equation, so the question is not yet settled. With this sort of definition, it would be difficult to establish whether or not a <u>given</u> equation was integrable. But one could try to <u>classify</u> all the integrable equations which arose from a certain type of linear system.

<u>References.</u>

Adler, M. & P. van Moerbeke 1980 Advances in Math. <u>38</u>, 267-379.

Athorne, C. & A.P.Fordy 1986 Generalized KdV equations associated with symmetric spaces. (Leeds preprint).

Chudnovsky, D. & G. 1984 Note on Eisenstein's system of differential equations. In: Classical and Quantum Models and Arithmetic Problems, eds. D. & G. Chudnovsky (Marcel Dekker), pp. 99-115.

Date, E., M. Kashiwara, M. Jimbo & T. Miwa 1983 Transformation groups for soliton equations. In: Proceedings of RIMS Symposium on Non-Linear Integrable Systems, eds. M. Jimbo & T. Miwa (World Scientific), pp. 39-119.

Dubois-Violette, M. 1982 Phys. Lett. B <u>119</u>, 157-161.

Eckhardt, B., J. Ford & F. Vivaldi 1984 Physica D <u>13</u>, 339-356.

Fordy, A.P. & P.P.Kulish 1983 Comm. Math. Phys. <u>89</u>, 427-443.

Forgacs, P. & N.S.Manton 1980 Comm. Math. Phys. <u>72</u>, 15-35.

Goldschmidt,Y.Y.& E.Witten 1980 Phys. Lett. B <u>91</u>, 392-396.

Olive, D. & N. Turok 1983 Nucl. Phys. B <u>215</u>, 470-494.

Penrose, R. 1976 Gen. Rel. Grav. <u>7</u>, 31-52.

Savvidy, G.K. 1984 Nucl. Phys. B <u>246</u>, 302-334.

van Moerbeke, P. 1985 Phil. Trans. R. Soc. Lond. A <u>315</u>, 379-390.

Ward, R.S. 1983 Gen. Rel. Grav. <u>15</u>, 105-109.

Ward, R.S. 1984 Phys. Lett. A <u>102</u>, 279-282.

Ward, R.S. 1984b Nucl. Phys. B <u>236</u>, 381-396.

Ward, R.S. 1985 Phys. Lett. A <u>112</u>, 3-5.

Wilson, G. 1985 Phil. Trans. R. Soc. Lond. A <u>315</u>, 393-404.

Witten, L. 1979 Phys. Rev. D <u>19</u>, 718-720.

MONOPOLE AND VORTEX SCATTERING

N.J. Hitchin
Mathematical Institute
24-29 St. Giles
Oxford
England.

§1.   This lecture is concerned with the Riemannian geometry which
arises by considering the motion of soliton-like solutions to the
Yang-Mills-Higgs equations.   These are the equations derived from
the action density

$$a = (F_A, F_A) + (D_A \phi, D_A \phi) + \lambda (1 - |\phi|^2)^2$$

in Minkowski space for a connection A with gauge group G and a
Higgs field ϕ in some representation of G.   In this expression
F_A is the curvature of A and $D_A \phi$ the covariant derivative of ϕ.
     The full Yang-Mills-Higgs equations may be considered as the
time evolution of a connection A and Higgs field ϕ on $\mathbb{R}^3$ and as
such they have the following mechanical interpretation.   We consider
the space of equivalence classes C of pairs (A, ϕ) under the action
of the group of gauge transformations G.   The tangent space at a
point on this infinite-dimensional manifold can be identified with
the infinitesimal deformations $(\dot{A}, \dot{\phi})$ orthogonal in L^2 to the gauge
orbits.   Then C acquires a Riemannian metric.

$$\frac{1}{2} \int_{\mathbb{R}^3} (\dot{A}, \dot{A}) + (\dot{\phi}, \dot{\phi}) .$$

     There is also a potential function defined on C by the gauge-
invariant quantity

$$V = \int_{\mathbb{R}^3} (F_A, F_A) + (D_A \phi, D_A \phi) + \lambda (1 - |\phi|^2)^2$$

associated to the connection and Higgs field on $\mathbb{R}^3$.
     The Yang-Mills-Higgs equations can then be viewed as the motion
of a particle on this infinite-dimensional configuration space with
kinetic energy term given by the metric and a potential V.
     Manton [10] took this interpretation further by suggesting that,
just as a ball bearing rolling inside a bowl will with small velocities
roll around the base of the bowl, so small velocity soliton-like

solutions to the Yang-Mills-Higgs system will approximate geodesic
motion on the absolute minimum of the potential function V. Clearly
there are analytical problems to be solved in order to make the argument rigorous but nevertheless it is one which leads us in a number
of cases to a rather special finite-dimensional Riemannian geometry
as we shall see.

§2.   There are two situations where the potential function V is
minimized in a known manner by particle-like configurations (A, ϕ).
The first is the 3-dimensional case of <u>magnetic monopoles</u>, where the
gauge group G is SU(2), the Higgs field is in the adjoint
representation, and $\lambda = 0$ in the potential term above. The second
is the 2-dimensional situation of <u>vortices</u>, where G = U(1), ϕ
lies in a complex one-dimensional representation and $\lambda = \frac{1}{4}$. These
cases are both described in the book of Jaffe and Taubes [9].

In both cases there is a topological invariant which provides a
value for the absolute minimum of V, given suitable boundary conditions. In the monopole case, if the Higgs field is assumed to
satisfy $|\phi| \to 1$ as $R \to \infty$, which is the last vestige of the quartic
potential term, then ϕ defines a map from a large sphere of radius R
to the unit sphere in the Lie algebra of SU(2) and the degree of
this map is the <u>magnetic charge</u> k. In the case of a vortex we also
have $|\phi| \to 1$ giving a map from a large circle in $\mathbb{R}^2$ to the unit
circle, and the degree k of this is again a topological invariant.
Moreover, as Jaffe and Taubes show, there are solutions to the
corresponding equations whose energy density is concentrated around k
points, providing at least for large separations, a particle-like
description.

The equations which realize this absolute minimum are first order
equations. For magnetic monopoles they are the <u>Bogomolny equations</u>:

$$F_A = {}^*D_A \phi$$

and for vortices the <u>vortex equations</u>:

$$\bar{\partial}_A \phi = 0 \; ; \qquad F_A = \frac{1}{2}(|\phi|^2 - 1) \; .$$

There is a difference of status between these equations at the present
time. Whereas, through a variety of methods, the Bogomolny equations

may be reduced to the algebraic geometry of a curve of genus $(k - 1)^2$, explicit solutions of the vortex equations are essentially unknown, although some analytical information is certainly available [11].

Nevertheless, the structure of the space of gauge equivalence classes is known.   For monopoles it is provided by Donaldson's description [3] as the space of rational maps

$$f(z) = \frac{a_0 + a_1 z + \ldots + a_{k-1} z^{k-1}}{b_0 + b_1 z + \ldots + b_{k-1} z^{k-1} + z^k}$$

of degree k (i.e. where denominator and numerator have no common factor), where a_i, b_j are complex numbers.   For vortices, the existence theorem of [9] shows that the space is the complex space of polynomials

$$p(z) = a_0 + a_1 z + \ldots + a_{k-1} z^{k-1} + z^k ,$$

a k-dimensional complex vector space.

The question which the scattering problem reduces to is to find the natural metric on each of these spaces which arises from the equations whose solutions they parametrize.

§3.   Fortunately, in the monopole case, this metric is of a very special kind - a hyperkähler metric.   This is a metric g which is compatible with three covariant constant complex structures I, J and K which satisfy the quaternionic identities:

$$I^2 = J^2 = K^2 = -1; \qquad IJ = - JI = K \quad \text{etc.}$$

Such metrics automatically have zero Ricci tensor, so they are solutions to the (positive definite) Einstein vacuum equations.   The complex structures give rise to three symplectic forms

$$\omega_1(X,Y) = g(IX,Y); \qquad \omega_2(X,Y) = g(JX,Y); \qquad \omega_3(X,Y) = g(KX,Y)$$

which lead us into the consideration of symplectic geometry.

In order to see why the monopole metric should be hyperkähler, recall the moment map in symplectic geometry: if a Lie group G acts symplectically on a symplectic manifold M with symplectic form ω, then the moment map $\mu : M \to g^*$ is an equivariant map to the dual of the Lie algebra of G such that

$$< d\mu, \xi > = i(X_\xi)\omega$$

where X_ξ is the vector field generated by $\xi \in g$.

Within symplectic geometry, if G acts freely on M, then the manifold

$$\mu^{-1}(0)/G$$

is again symplectic: the <u>Marsden-Weinstein quotient</u>. An analogous result holds in hyperkähler geometry [8]. If a Lie group G acts freely on a hyperkähler manifold, we obtain three moment maps μ_1, μ_2 and μ_3 corresponding to the symplectic forms ω_1, ω_2 and ω_3 and then

$$\bigcap_{i=1}^{3} \mu_i^{-1}(0)/G$$

is again hyperkähler.

The relevance of this result to magnetic monopoles is the following infinite-dimensional point of view. We consider the connection A and Higgs field ϕ on $\mathbb{R}^3$ as a vector in an infinite dimensional quaternionic vector space:

$$\phi + iA_1 + jA_2 + kA_3$$

which with the L^2 inner product is a flat hyperkähler manifold.

The group of gauge transformations G acts on this and the three moment maps turn out to be the three components of the Bogomolny equations. Thus

$$\bigcap_{i=1}^{3} \mu_i^{-1}(0)/G$$

is the space of equivalence classes of solutions to the Bogomolny equations, and its natural metric is hyperkähler.

§4. In the vortex case, life is not quite so simple. However, suppose we consider $SU(2)$ connections A on $\mathbb{R}^2$ and Higgs fields ϕ_1, ϕ_2 in the adjoint representation. Then writing

$$A_1 + iA_2 + j\phi_1 + k\phi_2$$

we have an infinite-dimensional hyperkähler manifold acted on by
gauge transformations in the same way as for monopoles.

Now for a symplectic manifold M we need not take simply
$\mu^{-1}(0)/G$ to create a new symplectic manifold, but we can consider
also $\mu^{-1}(x)/G_x$ where G_x is the stabilizer of $x \in g^*$. This again
is symplectic, and likewise the corresponding result in the hyperkähler situation. Taking

$$x = \begin{pmatrix} \tfrac{1}{2}i & 0 \\ 0 & -\tfrac{1}{2}i \end{pmatrix}$$

we obtain a hyperkähler metric on $\cap \mu_i^{-1}(x)/G_x$ which is the space of
equivalence classes solutions of the equations

$$\left. \begin{array}{l} \bar{\partial}_A \Phi = 0 \\[2mm] F_A + \tfrac{1}{2}[\Phi, \Phi^*] = x \end{array} \right\}$$

where ϕ_1 and ϕ_2 are incorporated into one complex Higgs field Φ.
Here equivalence is with regard to the gauge transformations leaving
x invariant i.e. U(1) gauge transformations.

Consider the special case where $\Phi = \begin{pmatrix} 0 & \phi \\ 0 & 0 \end{pmatrix}$ and the connection A reduces to a U(1) connection. Then the above equations
reduce to the vortex equations

$$\left. \begin{array}{l} \bar{\partial}_A \phi = 0 \\[2mm] F_A = \tfrac{1}{2}(|\phi|^2 - 1) \end{array} \right\} \quad .$$

An invariant way of characterizing them is the solutions (A, ϕ_1, ϕ_2)
which are invariant in the space of equivalence classes under the
circle action $\phi \rightarrow e^{i\theta}\phi$. In other words those for which this circle
action is generated by a gauge transformation.

This way the natural metric on the space of equivalence classes
of vortices is the restriction of a hyperkähler metric to the fixed
point set of a circle action [7].

§5. Reverting to the magnetic monopole situation, general information
about this metric may be derived from the methods of solving the
Bogomolny equations themselves [2]. In the particular case of k = 2
the hyperkähler structure itself is sufficient to lead to an explicit
solution. This is because, if we reduce to the situation of 2 monopoles which are centred about the origin, the space of equivalence
classes is 4-dimensional, hyperkähler and has an SO(3) action by
isometries arising from the rotation group of $\mathbb{R}^3$.

Such metrics were studied by Gibbons and Pope [5] and can be put
in the form:

$$ds^2 = w_1 w_2 w_3 d\eta^2 + \frac{w_2 w_3}{w_1} \sigma_1^2 + \frac{w_3 w_1}{w_2} \sigma_2^2 + \frac{w_1 w_2}{w_3} \sigma_3^2$$

where σ_1, σ_2 and σ_3 are a standard orthonomial basis of left-
invariant 1-forms on SO(3) and the functions w_1, w_2 and w_3
satisfy the equations:

$$w_1' + w_2' = - 2w_1 w_2 \; ; \qquad w_2' + w_3' = -2w_2 w_3$$

$$w_3' + w_1' = -2w_3 w_1 \; .$$

These equations were, in fact, solved by Halphen in 1881 [6] with
elliptic integrals. In this monopole situation the elliptic curve
which generates the solution of the Bogomolny equations yields a
reason for this.

Combined with the requirements of non-singularity and completeness one obtains a unique solution to the above problem and the metric
is completely determined [2].

As to the geodesic motion, the SO(3)-invariance generates two
constants of the motion, the metric a third but there has not yet
appeared a fourth commuting integral to make the flow completely
integrable. It is worth pointing out that although there are various
Killing tensors of higher rank on this manifold, none of them appears
capable of contraction to give such a fourth integral, analogous to
that which exists for the Kerr solution in general relativity.

Nevertheless, the motion on some totally geodesic surfaces of
revolution may be analyzed [1], [2] to give interesting scattering
behaviour of the monopoles thought of as particles.

§6.   In the case of k = 2 for centred <u>vortices</u>, the space of
equivalence classes is $\mathbb{R}^2$, but this lies, as described in §4, as
the fixed point set of a circle action on a 4-dimensional hyperkähler
manifold M.

In this case there is no SO(3) action, but there is a circle
action arising from physical rotations in $\mathbb{R}^2$ and the circle action
$\phi \rightarrow e^{i\theta}\phi$ which is an internal or gauge action.   Some combination of
these two actions must leave invariant the symplectic forms ω_1, ω_2
and ω_3 to yield moment maps μ_1, μ_2, μ_3, and a map $\mu : M \rightarrow \mathbb{R}^3$.
This represents M as a circle bundle (outside the fixed points of
the action) over $\mathbb{R}^3$ and leads to the following ansatz [4] for such
a metric:

$$ds^2 = V d\underline{x} \cdot d\underline{x} + V^{-1}(d\tau + \underline{\omega})^2$$

where grad V = curl $\underline{\omega}$.

The metric itself therefore depends just on the harmonic function
V, which because of the spatial rotation invariance is axially
symmetric on $\mathbb{R}^3$.   Beyond this, we have little data to determine V
either because we do not know how to solve the vortex equations
explicitly or because we do not have enough qualitative information
about the generalized SU(2) vortex equations introduced in §4 to
impose boundary conditions on V to determine it.

We cannot say very much, then, about vortex scattering but there
is one feature which both monopoles and vortices share, which is a
consequence of the existence of the metric and the smoothness of the
parameter space.   This is the phenomenon of 90° scattering whereby
a geodesic passing through an axially symmetric solution leads to the
picture of two particles undergoing a direct collision, and subsequently
splitting into two particles which move off in directions orthogonal
to the original ones.   This is something which can be seen directly
for monopoles as in [1] and has been studied by P. Ruback in the case
of vortices.

References

[1]   M.F. Atiyah & N.J. Hitchin, Low energy scattering of non-
      abelian monopoles, Phys. Lett. <u>107A</u> (1985), 21-25.

[2] M.F. Atiyah & N.J. Hitchin, "The geometry and dynamics of
     magnetic monopoles", Princeton University Press (to appear).

[3] S.K. Donaldson, Nahm's equations and the classification of
     monopoles, Commun. Math. Phys. $\underline{96}$ (1984), 387-407.

[4] G. Gibbons & S.W. Hawking, Gravitational multiinstantons,
     Phys. Lett. $\underline{B78}$ (1978), 430-432.

[5] G. Gibbons & C. Pope, The positive action conjecture and
     asymptotically Euclidean metrics in quantum gravity, Commun.
     Math. Phys. $\underline{66}$ (1979), 267-290.

[6] G.H.Halphen, Sur un systeme d'equations differentielles, C.R.
     Acad. Sci. Paris $\underline{92}$ (1881), 1101-1103.

[7] N.J. Hitchin, Metrics on moduli spaces, in "Contemporary
     Mathematics", Volume $\underline{58}$, Part I (1986), American Mathematical
     Society, Providence.

[8] N.J. Hitchin, A Karlhede, U. Lindström & M. Roček, Hyperkähler
     metrics and supersymmetry, Commun. Math. Phys. (to appear).

[9] A. Jaffe & C.H. Taubes, "Vortices and monopoles", Birkhauser,
     Boston (1980).

[10] N.S. Manton, A remark on the scattering of BPS monopoles, Phys.
     Lett. $\underline{110}$ B (1982), 54-56.

[11] H.J. de Vega & F.A. Schaposnik, Phys. Rev. $\underline{D14}$ (1976), 1100.

The Ambitwistor Program

James Isenberg
Dept. of Mathematics
University of Oregon
Eugene, OR   97403
USA

The goal of the ambitwistor program is to obtain a representation
of the fundamental physical fields in terms of geometrical
constructions on ambitwistor spaces, which are spaces whose points
correspond to null geodesics in spacetime. The hope is that in this
new formulation, the fields will become accessible to new techniques
of analysis, and perhaps one will thus be led to new insights, both
physical and mathematical. In this brief sketch, we discuss some of
the successes which the program has achieved to date, and indicate
some of the directions in which current research is heading.

The ambitwistor program is of course an outgrowth of the twistor
programme of Penrose.[1] Indeed, many of the ideas and techniques
arising in ambitwistorial analyses parallel those used in
corresponding twistorial analyses. In a broad sense, both programs
are based on two major themes: (1) Focus on conformally invariant
geometric structures in spacetime (e.g., null planes and null
geodesics), (2) Rely as much as possible on holomorphic (i.e. complex
analytic) mathematical analysis. More particularly, as we shall see,
in many cases the ambitwistorial representation of certain fields
closely parallels the twistorial representation.

However, while working with twistors usually restricts one to the
representation of self-dual and antiself-dual fields, the use of
ambitwistor spaces allows one to represent general (non self-dual)
fields as well. Ambitwistor spaces also seem to more readily handle
fields with supersymmetry than do twistor spaces. Hence the
ambitwistor approach has a potentially wider application. Currently,
it should be noted, ambitwistor analyses are more difficult to carry
out. But the hope is that if mathematicians begin to examine more
carefully the basic ambitwistorial spaces, these analyses will become
simpler and more practical.

Before discussing how to use ambitwistorial techniques to
represent fields, we shall briefly describe (in Sec. 1) the basic
spaces themselves. Then in Sec. 2, we shall discuss the major success
of the program--the representation of Yang-Mills fields. We follow

this (in Sec 3) by considering gravity, which is presently a focus at research. Finally in Sec. 4, we comment on a possible role for ambitwistorial techniques in the study of strings and superstrings.

1. The Spaces

Twistor space itself is $T = \mathbb{C}^4$, with a specified action of each of the groups $SL(4,\mathbb{C})$ and $SU(2,2)$ on T. $SL(4,\mathbb{C})$ is the 4-fold cover of the conformal group of (compactified) complex Minkowski spacetime[2] $\mathbb{C}M$, while $SU(2,2)$ is the 4-fold cover of the conformal group of (compactified) real Minkowski spacetime M (sitting inside $\mathbb{C}M$). Based on T, one defines[3]

Projective Twistor Space:

    PT = {twistors, modulo complex conformal scale factors}
      $\cong P^3(\mathbb{C})$

Dual Twistor Space:

    $T^* = $ {linear maps $W:T \to \mathbb{C}:Z \mapsto \langle W|Z\rangle$}
      $\cong \mathbb{C}^4$

Projective Dual Twistor Space:

    $PT^* = $ {dual twistors, modulo complex conformal scale factors}
      $\cong P^3(\mathbb{C})$

Ambitwistor Space:

    $A = $ {$(Z,W) \in T \times T^*$ such that $\langle W|Z\rangle = 0$}
      $\cong$ 7 dimensional quadric in $\mathbb{C}^4 \times \mathbb{C}^4$

Projective Ambitwistor Space

    $PA = $ {$([Z],[W]) \in PT \times PT^*$ such that $\langle [W]|[Z]\rangle = 0$}
      $\cong$ 5 dimensional quadric in $P^3(\mathbb{C}) \times P^3(\mathbb{C})$

One of the keys to the utility of these various spaces in describing fields in spacetime is the correspondence between geometric objects in $\mathbb{C}M$ and geometric objects in these spaces. These results are easily derived algebraically[4] from a certain spinorial equation-- it appears in the literature in the form $ix^{AA'}\pi_{A'} = \omega^A$ -- relating points $x^{AA'}$ in $\mathbb{C}M$ and twistors

$Z = \begin{pmatrix} \pi_{A'} \\ \omega^A \end{pmatrix}$. We sketch the important ones in the following table:

CM	PT	PT*	PA
point	projective line [$P^1(C)$]	projective line [$P^1(C)$]	projective line [$P^1(C)$] x [$P^1(C)$]
null geodesic line	-	-	point
self-dual null plane ("β-plane")	projective plane [$P^2(C)$]	point	projective plane [$P^2(C)$]
anti-self-dual plane ("α-plane")	point	projective plane [$P^2(C)$]	projective plane [$P^2(C)$]

In this table, we find the key result that PA ≈ {null geodesics in ℂM}
{Note that for convenience, here and throughout this paper, we ignore
the subtleties regarding points and lines "at ∞".[1] It should be
pointed out, however, that for most applications (such as for the
correspondences described below) one works not with PT ≅ $P^3(ℂ)$ but
rather with PT ≅ $P^3(ℂ) - P^1(ℂ)$, or some subspace thereof. Such spaces
are generally not compact. Similarly, we shall (ambiguously) use "ℂM"
to denote all of, or portions (noncompact), of compactified
complexified Minkokski space.} In all of these spaces, the important
structures are complex analytic, rather than Hermitian.

2. Twistor and Ambitwistor Representations of Yang-Mills
Fields

Perhaps the most important success thus far of the twistor
programme (as well of its ambitwistor offshoot) is the representation
of Yang-Mills fields. Yang-Mills fields, we recall, are G-connection
fields A on spacetime which satisy the Yang-Mills field equations
D✱F = 0, where F is the curvature of A, D is the exterior covariant
derivative based on the connection A, ✱ is the Hodge dual, and G is
some Lie group. When a connection field A is self-dual (satisfying
the condition ✱F = F) or antiself-dual (satisfying ✱F = -F) then it
follows from the Bianchi identity (DF = 0) that the Yang-Mills
equation is automatically satisfied. Hence self-dual and antiself-
dual connections form a special subclass of the Yang-Mills solutions.
For these, one has the basic Ward Theorem[5] which establishes the

following one-one correspondences:

$$
\left\{
\begin{array}{c}
\text{Self-dual Connections A} \\
\text{on } \mathbb{C}M
\end{array}
\right\}
\longleftrightarrow
\left\{
\begin{array}{c}
\mathbb{C}^n \text{ Vector Bundles E} \\
\text{over PT} \\
[\text{Trivial on all P1}(\mathbb{C})\text{ subspaces}]
\end{array}
\right\}
\qquad (1)
$$

and

$$
\left\{
\begin{array}{c}
\text{AntiSelf-dual Connections A} \\
\text{on } \mathbb{C}M
\end{array}
\right\}
\longleftrightarrow
\left\{
\begin{array}{c}
\mathbb{C}^n \text{ Vector Bundles E} \\
\text{over PT*} \\
[\text{Trivial on all P1}(\mathbb{C})\text{ subspaces}]
\end{array}
\right\}
\qquad (2)
$$

Note that the dimension n of these vector bundles E corresponds to that of the Lie algebra in which the Yang-Mills fields take values.

These correspondences, which rely strongly on properties of holomorphic functions, generally involve connections which take values in complex groups--e.g., $GL(n,\mathbb{C})$. Remarkably, however, the condition that the connection be real (e.g., taking values in $GL(n,\mathbb{R})$, or in $SU(n)$) on an S^4 subspace of $\mathbb{C}M$ can be readily incorporated into the twistor side of these correspondences. In particular, if the bundle E admits a symplectic structure $\sigma: E \to E$ (with $\sigma^2 = -1$) which corresponds to a lift of the reality structure ρ on PT (that map ρ which determines real and imaginary parts of PT), then the corresponding (self-dual) Yang-Mills connection is real on S^4. Schematically, the result[6] takes the form

$$
\left\{
\begin{array}{c}
\text{Self-dual SU(n) Connections A} \\
\text{on } S^4 \\
\text{with instanton number k}
\end{array}
\right\}
\longleftrightarrow
\left\{
\begin{array}{c}
\mathbb{C}^n \text{ Vector Bundles E} \\
\text{over PT*} \\
[\text{Trivial on all P}^1(\mathbb{C})\text{ subspaces}] \\
\text{with symplectic } \sigma \\
\text{and second Chern number k}
\end{array}
\right\}
\qquad (3)
$$

As remarkable as this correspondence, is the fact that the $\mathbb{C}^n$ vector bundles E described on the right hand side of (3) are well-understood.[7] Consequently, the special Yang-Mills fields to which they correspond are also well understood, and can be explicitly constructed. These are the Yang-Mills instantons (on S^4).

Instantons and other self-dual Yang-Mills connections have turned out to be very important in mathematics,[8] and may play a role in quantum field theory calculations as well.[9] However, one wishes to also understand Yang-Mills fields which have no such (self-dual or antiself-dual) restriction. Ambitwistor space was introduced to aid in the study of such unrestricted Yang-Mills connections. One finds that if one considers holomorphic vector bundles over PA analogous to those over PT and PT* as in (1) and (2), then again one has a representation of GL(n) connections over complex Minkowski space. But now they are in general neither self-dual nor antiself-dual; nor do they generally solve the Yang-Mills equations. Schematically, one has[10]

$$\left\{ \begin{array}{c} \text{All Connections A} \\ \text{over } \mathbb{C}M \end{array} \right\} \longleftrightarrow \left\{ \begin{array}{c} \mathbb{C}^n \text{ Vector Bundles } \mathcal{I} \\ \text{over PA} \\ [\text{Trivial on all } P^1(\mathbb{C}) \times P^1(\mathbb{C}) \text{ subspaces}] \end{array} \right\} \quad (4)$$

To understand why the twistorial bundles correspond to self-dual connections while the ambitwistorial bundles correspond to connections with no such restriction, note that self-dual connections on $\mathbb{C}M$ are exactly those which are flat on all antiself-dual null planes in $\mathbb{C}M$. Now in verifying the correspondences (1), (2), and (4), one finds that the connections which correspond to bundles over a given twistor-type space (e.g., PT, PT*, or PA) are necessarily flat over the geometric structures in $\mathbb{C}M$ which correspond to points in that twistor-type space. Hence, recalling the table in Section 1, we see that bundles over PT correspond to connections flat over antiself-dual null planes (therefore the connections are self-dual), while bundles over PA determine connections which are flat over null lines (therefore no restriction).

As noted above, besides being free of any self-dual restriction, the connections corresponding to bundles over PA needn't satisfy the Yang-Mills equations. How do we build the Yang-Mills equations into the ambitwistorial representation of connections? Recall that PA sits as a quadric (codimension-one subset determined by an algebraic condition) in PT x PT*. Among the collection of all bundles over PA, there are some which extend to bundles over PT x PT*. The connections corresponding to these, one finds, satisfy the Yang-Mills equations. But in addition, they satisfy a number of other conditions which are generally too restrictive to be of interest. [If, for example, an SU(2) connection satisfies these conditions, it is necessarily either self-dual, antiself-dual, or Abelian]. To obtain conections which satisfy the Yang-Mills equations and nothing stronger, we look at bundles over PA which are extendible to PT x PT* only to third order in Taylor series. That is, the transition functions which define the bundle over PA may be expanded in a Taylor series involving a parameter "s" transverse to PA $\hookrightarrow$ PT x PT* ; but only to order s^3 are the expansion terms consistent with the cocycle conditions being satisfied in a neighborhood of PA in PT x PT*. In schematic form, we have[10,11]

$$
\left\{ \begin{array}{c} \text{Yang-Mills} \\ \text{Connections A} \\ \text{on } \mathbb{C}M \end{array} \right\} \longleftrightarrow \left\{ \begin{array}{c} \mathbb{C}^n \text{ Vector Bundles } \mathcal{F} \\ \text{over PA} \\ \text{[Trivial on all P1(}\mathbb{C}\text{) x P1(}\mathbb{C}\text{) subspaces]} \\ \text{[3rd order extendible to PT x PT*]} \end{array} \right\} \quad (5)
$$

This correspondence could be as useful in studying nonself-dual Yang-Mills solutions as has been the Atiyah-Ward representation of self-dual Yang-Mills fields. This has not been the case, however, because PA is not as familiar a space to mathematicians as is PT[$\cong$ P^3($\mathbb{C}$) $-$ P^1($\mathbb{C}$)].

We wish to briefly describe two other representations of Yang-Mills connections which are closely related to the ambitwistorial one just discussed. The first of these, which appears as an intermediate step of our proof of correspondence (5)[10], eschews twistor-like spaces and focuses on complex Minkowski space $\mathbb{C}M$ embedded as a 4 complex dimensional diagonal in $\mathbb{C}M^8$ (which is just $\mathbb{C}^8$ with an orthogonal metric). Let us assume that an orthogonal split of $\mathbb{C}M^8$ has been chosen so that we have $\mathbb{C}M^8 = \mathbb{C}M_L^4 \times \mathbb{C}M_R^4$ ["Physical" complex Minkowski space, $\mathbb{C}M$, is everywhere transverse to this split, as in the diagram]. We define a connection ^{8}A on $\mathbb{C}M^8$ to be <u>bidual</u> if its curvature ^{8}F satisfies the conditions

$^8F_{\overleftrightarrow{ab}}$ is self-dual (where "$\overleftrightarrow{a}$" indicates restriction to vectors in $T\mathbb{C}M_L^4$))

$^8F_{\overrightarrow{ab}}$ is antiself-dual (where "$\overrightarrow{a}$" indicates restriction to vectors in TCM_R^4)

$^8F_{\overrightarrow{ab}} = 0$

We then discover that a connection ^{4}A on the diagonal spacetime $\mathbb{C}M$ satisfies the Yang-Mills equations on the diagonal if and only if this connection is extendible to a connection ^{8}A on $\mathbb{C}M^8$, and that to first order in a parameter transverse to $\mathbb{C}M$, the connection ^{8}A is bidual.

This $\mathbb{C}M \hookrightarrow \mathbb{C}M^8$ representation of Yang-Mills fields motivates some of our schemes for trying to find an ambitwistor representation of gravitational field equations, as discussed below (in Section 3).

Another representation of Yang-Mills fields, closely related to the ambitwistorial correspondence (5), is that which Witten has developed using a supersymmetric version of PA. Recall that one may regard standard projective ambitwistor space as the collection of null geodesics in complex Minkowski space $\mathbb{C}M$. These geodesics are the stationary points of the action $\int d\tau \; (g_{\mu\nu} \frac{dx^\mu}{d\tau} \frac{dx^\nu}{d\tau})$ for the motion of a massless free particle with trajectory $x^\mu(\tau)$ in $\mathbb{C}M$ with orthogonal metric $g_{\mu\nu}$. Based on this interpretation of PA, one may define (following Witten[11]) a generalized ambitwistor space $PA_{[n,s]}$ consisting of the trajectories of free massless supersymmetric particles moving in a super (complex) Minkowski space $\mathbb{C}M_{[n,s]}$ with n commuting dimensions and s super ones.

Witten has studied two such spaces--$PA_{[4,12]}$ and $PA_{[10,16]}$--and found that they both lead to interesting representations of Yang-Mills fields.[11,12] Appropriate vector bundles over $PA_{[4,12]}$ correspond to N = 3 super Yang-Mills solutions on $\mathbb{C}M_{[4,12]}$, while appropriate bundles over $PA_{[10,16]}$ correspond to N = 1 super Yang-Mills solutions on $\mathbb{C}M_{[10,16]}$. The first is noteworthy because the N = 3 super Yang-Mills connections induce standard Yang-Mills connections (generally nonself-dual) on $\mathbb{C}M^4$. So we have a representation of Yang-Mills fields which avoids any explicit stipulations of bundle extendibility. The second is noteworthy because the recent popularity of superstring theory has led to interest in superspacetimes with 10 real dimensions.

How do these superambitwistorial representations of Yang-Mills fields avoid bundle extendibility requirements? Recall, again, that in these twistor-type representations of connections, the construction guarantees that the connections will be flat over the structures in spacetime which correspond to points in the twistor-type space. Now the points in $PA_{[n,s]}$ (for $s \neq 0$), though often referred to as "super null lines", are multidimensional objects (one nonsuperdimension plus a number of superdimensions). The requirement that a connection be flat on these is therefore nonvacuous. In the cases mentioned above-- $PA_{[4,12]}$ and $PA_{[10,16]}$--this requirement imposes the Yang-Mills field equations. It should be noted that, in a certain sense, the extendibility condition is smuggled into the superambitwistor representation. One sees this when one relates bundles over supermanifolds to bundles over ordinary manifolds.

3. Twistor and Ambitwistor Representation of Gravitational Fields

The geometry of complex Minkowski space is built into the structure of projective twistor space PT and projective ambitwistor space PA. Hence PT and PA are not themselves useful for representing general (curved) spacetimes. As shown by Penrose, however, deformations[13] of these spaces are in fact useful for this task. Specifically, he shows[14] that deformations of PT which preserve its fibration over $P^1(\mathbb{C})$ and also preserve a certain (deformed) vertical two form μ can be used to represent self-dual spacetimes which are deformations of complex Minkowski space. [A self-dual spacetime is one for which the Ricci curvature vanishes and the Weyl curvature is self-dual]. Schematically, this correspondence may be written as

$$
\left\{\begin{array}{c} \text{Selfdual Spacetime} \\ (\mathfrak{m}^4, g) \\ \text{[Deformation of CM]} \end{array}\right\} \longleftrightarrow \left\{\begin{array}{c} \text{Bundles } \mathcal{P} \\ P^1(\mathbb{C}) \\ \text{with vertical 2-form } \mu \\ \text{[Deformation of PT with } \mu] \\ P^1(\mathbb{C}) \end{array}\right\} \tag{6}
$$

These self-dual spacetimes, called "nonlinear gravitons" by Penrose, admit flat self-dual nullplanes [referred to in the literature as "α-planes"]. Indeed, for a given nonlinear graviton spacetime, the corresponding space $\mathcal{P}$ is essentially the collection of these α-planes, and some of the geometric correspondences which hold between PT and $\mathbb{C}$M survive: For example, the preserved fibration of $\mathcal{P}$ over $P^1(\mathbb{C})$ ties together (in fibres) all parallel α-planes in $\mathfrak{m}^4$; and the sections of this fibration correspond to points in $\mathfrak{m}^4$. On the other hand, there do not generally exist flat antiself-dual null planes ("β-planes") in $\mathfrak{m}^4$, and this is reflected in the structure of any nontrivially deformed $\mathcal{P}$. Note that, roughly speaking, μ carries the information regarding the conformal scale of the spacetime $(\mathfrak{m}^4, g)$.

In a spacetime with nonself-dual curvature, there are generally no α or β null planes. However there are always geodesic curves, and so one is led to consider ambitwistorlike spaces for representing general curved spacetimes. Yasskin and I[15] have chosen to consider deformations of ambitwistor space which, like $\mathcal{P}$, preserve the characteristic foliation (for PA, this foliation is over $P^1(\mathbb{C})$ x $P^1(\mathbb{C})$. Our deformations also preserve a certain collection of "contact forms", which we label as $\{\sigma_p\}$. We get

$$\left\{\begin{array}{c} \text{Teleparallel Spacetimes} \\ (\mathfrak{M}^4,g,\mathbf{v}) \\ \text{[Deformations of CM]} \end{array}\right\} \longleftrightarrow \left\{\begin{array}{c} \text{Bundles } \mathcal{A} \\ P^1(\mathbb{C}) \times P^1(\mathbb{C}) \\ \text{with contact forms } \{\sigma_p\} \\ \text{[Deformations of PA with } \sigma_p] \\ \downarrow \\ P^1(\mathbb{C}) \times P^1(\mathbb{C}) \end{array}\right\}$$

Note that a teleparallel spacetime is one which has a metric compatible connection whose curvature vanishes, but whose torsion generally does not vanish.

We seem to be guilty of false advertising here: We claimed to have a correspondence for the configuration space of all spacetime geometries (presumed torsion-free); instead our correspondence is for spacetimes with teleparallel geometries. But in fact any (parallelizable) torsion free spacetime geometry ($\mathfrak{M}^4$,g) with arbitrary Riemann curvature can be represented by a teleparallel geometry ($\mathfrak{M}^4$,g,$\mathbf{v}$): One simply chooses an orthonormal frame field $\{e_\alpha\}$ for $\{\mathfrak{M}^4$,g$\}$ and then defines $\mathbf{v}$ to be that connection for which $\{e_\alpha\}$ is parallel. The relation between the torsion of $\mathbf{v}$ and the curvature of g [more properly, the curvature R of the Levi–Civita connection for g] is then given by

$$R^m_{[g]nab} = \mathbf{v}_a k^m{}_{nb} - \mathbf{v}_b k^m{}_{na} + k^m{}_{pb} k^p{}_{na} - k^m{}_{pa} k^p{}_{nb}$$

where $k^a{}_{bm} = \frac{1}{2}(Q_{mb}{}^a + Q_{bm}{}^a - Q^a{}_{mb})$, and $Q^a{}_{bm}$ is the torsion. Going the other way, a teleparallel geometry determines a torsion-free geometry: One simply forgets $\mathbf{v}$ and works with the Levi–Civita connection (metric-compatible, torsion-free). The relationship is many-to-one, with the class of teleparallel geometries for a given torsion-free geometry parametrized by the set of orthonormal frame fields for the given metric. This is the "gauge freedom" of the correspondence {spacetimes ($\mathfrak{M}^4$,g)} $\longleftrightarrow$ {$\mathcal{A}$, {σ_p}}.

Why bother with teleparallel connections? From our perspective, they are an artifact of a well-motivated and well behaved class of deformations of ambitwistor space, and we believe that they will not get in the way of our obtaining a Penrose-type correspondence for solutions of Einstein's equations. However, one may consider other deformations of ambitwistor space which avoid teleparallel geometries. This is the program of LeBrun, Eastwood, Baston, and Mason.[16] They consider general deformations of PA which preserve a global contact form but generally lose the fibration. Their global contact form guarantees no torsion. Our fibration guarantees teleparallelism.

(Each fibre is a collection of parallel null geodesics).

As of yet, there is no firm result concerning the incorporation of the Einstein equations (or any alternative set of gravitational field equations) into the structure of deformed ambitwistor spaces (either of the sort considered by Yasskin and me, or the sort studied by LeBrun, Eastwood, Baston, and Mason). The Yang-Mills results have led us to a plausible conjecture, however. The idea is to consider deformations of $PT \times PT^*$ which preserve the fibration over $P^1(\mathbb{C}) \times P^1(\mathbb{C})$ along with certain other structures (analogous to the two-form μ preserved in the nonlinear graviton $\mathcal{P}$.) Let us call these spaces generically $\mathcal{P}^{\alpha\beta}$. Then, one conjectures that a given spacetime will satisfy the gravitational field equations if and only if the corresponding deformed ambitwistor space $\mathcal{A}$ is embeddible--perhaps only to certain order--in some $\mathcal{P}^{\alpha\beta}$.

Testing this conjecture directly is formidable. Hence we have chosen to study it indirectly, via an intermediary. Recall that a G-connection on complex Minkowski space satisfies the Yang-Mills equations if and only if when this connection is extended into $\mathbb{C}M^8$, it is bidual to second order in the extension parameter [see section 2]. Now let us consider a spacetime $(\mathcal{M}^4, g, \triangledown)$ and assume it embeds into some eight complex dimensional space $\mathcal{M}^8$ with metric ^8g and connection $^8\triangledown$. We call such a spacetime $(\mathcal{M}^8, {}^8g, {}^8\triangledown)$ "bidual" if and only if (1) it admits a pair of four complex dimensional distributions $T_L\mathcal{M}^8$ and $T_R\mathcal{M}^8$ (not necessarily integrable) which split the tangent space at each point; and (2) the metric ^8g and connection $^8\triangledown$ split orthogonally relative $T_L\mathcal{M}^8$ and $T_R\mathcal{M}^8$; with the curvature and torsion being self-dual relative to $T_L\mathcal{M}^8$ and antiself-dual relative to $T_R\mathcal{M}^8$ at each point. (These conditions are described more precisely elsewhere[17]). Requiring that $(\mathcal{M}^4, g, \triangledown)$ embed in a bidual spacetime is presumably a very strong condition. Our conjecture is that the appropriate requirement is that $(\mathcal{M}^4, g, \triangledown)$ embeds in some $(\mathcal{M}^8, {}^8g, {}^8\triangledown)$ which is bidual only (to some order) in a neighborhood of $(\mathcal{M}^4, g, \triangledown)$. We also conjecture that such embeddibility occurs if and only if the corresponding $\mathcal{A}$ can be embedded (to some corresponding order) in some $\mathcal{P}^{\alpha\beta}$. Note that Yasskin and I have proven that there is a one-one correspondence between bitwistor spaces $\mathcal{P}^{\alpha\beta}$ and bidual spacetimes $(\mathcal{M}^8, T_L\mathcal{M}^8, T_R\mathcal{M}^8, {}^8g, {}^8\triangledown)$[15].

4. Ambitwistors and Strings

To date, the suspicion that ambitwistorial (or twistorial) techniques might be of some use in studying superstring theory is

largely based upon two sets of interesting results: 1) As shown by
Witten[12] (and as noted above), there is a well-defined ambitwistor
space $PA_{[10,16]}$ corresponding to super Minkowski space with ten real
and sixteen superdimensions; and one can use bundles over $PA_{[10,16]}$ to
represent connections satisfying the super Yang-Mills equations over
super Minkowski space. 2) Developing and extending work by
Weirstrauss[19], Shaw[20] has shown that solutions of the classical
string field equations [i.e., minimal surfaces] in three, four, six, and
ten dimensions may be represented by pairs of curves in generalized
twistor spaces.

Either of these results may well lead to fruitful studies of
superstring theories.[21] The approach I prefer, however involves a
different space. Recalling the definition of $PA_{[n,s]}$ in terms of null
geodesic trajectories of massless particles in super Minkowski space,
we define

$$B_{[n,s]} := \{\text{world sheets of strings in } CM_{[n,s]}\}$$

[A similar space may be defined for worldsheets of strings in nonflat
spacetimes.] The properties of these infinite dimensional spaces are
far from understood. The hope is that in studying their holomorphic
structure, and perhaps also in examning certain holomorphic bundles
over $B_{[n,s]}$, one might obtain some insight regarding either the full
(all-energy) superstring theory, or the low energy theory (manifest as
gravitational and Yang-Mills fields). The former study could involve
studying appropriately defined "string instantons". The latter may
involve establishing relationships between $B_{[n,s]}$ and $PA_{[n,s]}$.

This is of course just speculaton. I believe, however, that
ambitwistorial ideas could play an interesting role in the development
of superstring theory.

Acknowledgements:

I thank Professors N. Sanchez and H. deVega for inviting me to
speak at their seminar in Meudon.

References

1.   Comprehensive reviews of twistor theory appear in: Penrose, R. and
     Ward, R.S., "Twistors for Flat and Curved Spacetime" in _General
     Relativity and Gravitation_ (ed. A. Held), Plenum, 1980;
     Hughston, L.P. and Ward, R.S., _Advances in Twistor Theory_,
     Pitman, 1979; Wells, R.O., _Complex Geometry in Mathematical
     Physics_, SMS #78, Les Presses de l'Univ. de Montreal, 1982;
     Penrose, R., and Rindler, W., _Spinors and Space-time_ (2 volumes)

Cambridge, 1984-1985; Hugget, S.A., and Tod, K.P., _An Introduction to Twistor Theory_, Cambridge, 1985.

2. For a discussion of $\mathbb{C}M$, see Penrose and Rindler, ref. [1].

3. These spaces, as well as others which play an important role in the twistor programme, are all "flag manifolds" of $T = \mathbb{C}^4$. See Wells, R., _Complex Geometry in Mathematical Physics_, Les Presses de L'Universite de Montreal, 1982.

4. See references above, or Newman, E.T. and Hansen, R., Gen. Rel, Grav. _6_, 361, 1975.

5. Ward, R., Phys. Lett. _A61_, 81, 1977.

6. Atiyah, M., and Ward, R., Comm. Math. Phys. _59_, 117, 1977.

7. Atiyah, M.F., Drinfeld, V.G., Hitchin, N.J., and Manin, Yu. I., Phys. Lett. _A65_, 185, 1978. Atiyah, M.F., _Geometry of Yang-Mills Fields_, Scuola Normale Superiore, 1979.

8. We refer especially to the work on "Fake R^4s". See Donaldson, S., J. Diff. Geom. _18_, 269, 1983; Freed, D., and Uhlenbeck, K., _Instantons and Four-Manifolds_, Springer, 1984.

9. Coleman, S., "The Uses of Instantons", lecture notes from 1977 International School of Subnuclear Physics, 1977.

10. Isenberg, J., Yasskin, P.B., and Green, P., Phys. Lett. _78B_, 462, 1978. Isenberg, J. and Yasskin P.B., "Twistor Description of Non-Self-Dual Yang-Mills Fields, in _Complex Manifold Techniques in Theoretical Physics_ (eds. D. Lerner and P. Sommers), Pitman, 1979.

11. Witten, E., Phys. Lett. _77B_, 394, 1978.

12. Witten, E., Nucl. Phys., _B266_, 245, 1986.

13. A "deformation" of a complex manifold may be understood as a smoothly parametrized set of transformations of the complex transition functions (on patch overlaps) which define the holomorphic structure of the manifold. The paramtrized set must include the identity transformation.

14. Penrose, R., Gen. Rel. Grav. _7_, 31, 1976. Curtis, W.D., Lerner, P.F., and Miller, F.R., Gen. Rel. Grav. _10_, 557, 1976. Hitchin, N.J., Math. Proc. Camb. Phil. Soc. _85_, 465, 1979.

15. Yasskin, P.B., and Isenberg, J., Gen. Rel. Grav. _14_, 621, 1982.

16. LeBrun, C., Trans AMS _278_, 208 1983. Eastwood, M., Twistor Newsletter 17, 1983. Basten, R., and Mason, L., Twistor Newsletter 21, 1986.

17. Isenberg, J., and Yasskin, P., Twistor Newsletter 22, 1986.

18. Unpublished.

19. Weirstrauss, K., Monats, Berliner Akad., 612, 1866.

20. Shaw, W., Class and Qtm Grav 2, 113, 1985. Shaw, W., to appear in _Mathematics in General Relativity_, (ed., J. Isenberg) AMS Contemp Math.

21. It would be interesting if someone were to apply some of Segal's ideas on quantum field to the space of classical string field solutions which arises in Shaw's work. See Segal, I., J. Math. Phys. 1, 468, 1959.

SUPERSYMMETRIC EXTENSION OF TWISTOR FORMALISM

J. Lukierski[*]
International Centre for Theoretical Physics,
34100 Trieste, Italy

Contents

1. Introduction

In the twistor programme of replacing D=4 Minkowski space-time geometry by more elementary complex geometry of twistors (see [1]) one can
distinguish two approaches:

a) one introduces totally null <u>2-planes</u> as primary geometric objects
in complex Minkowski space (CM). These null-planes in CM, called α-pla-
nes, are described by single projective twistors (PT), i.e.

$$\text{point in PT} \quad \leftrightarrow \quad \alpha\text{-plane in CM} \qquad\qquad (A)$$

If we hold the point in CM, and vary the (nonprojective) twistors (T),
we get the linear 2-space in T, i.e.

$$\begin{array}{c}\text{2-plane in T}\\ \text{(line in PT)}\end{array} \quad \leftrightarrow \quad \text{point in CM} \qquad\qquad (B)$$

The correspondence (A) describes the twistor transform. It appears that

i) The description of YM equations in PT implies the integrability
conditions on α-planes, which select only self-dual (instanton) solutions (see e.g. [2-3]).

---

*)On leave of absence from the Institute for Theoretical Physics, Uni-
versity of Wroclaw, ul.Cybulskiego 36, 50-205 Wroclaw, Poland

ii) the generalization of the notion of α-planes from flat space to general 4-dimensional complex manifold M imposes severe restrictions on its conformal geometry: the Weyl tensor should be essentially self--dual. However, there were efforts to generalize the class of complex-analytic deformations of α-planes[1], none were able to describe general solution of Einstein equations.

Therefore in order to be able to describe existing field theories

b) one introduces null <u>lines</u> in CM as primary geometric objects. If we interpret $z_A \in T$ (A=1...4) as conformal spinors, one can introduce the SU(2,2) - invariant scalar product $<z,z'>$, and define the <u>ambitwistor space</u> as follows:

Projective Ambitwistor space: $PA = (z,w) \in PT \times \overline{PT}$

$$<z,w> = 0 \qquad\qquad (1.1)$$

where by $w_A \subset \overline{T}$ we denoted dual twistors defining linear maps T→C : z→<z,w> and respectively projective dual twistors $(\overline{PT})$ are defined modulo complex scale factor. One can show that

point in PA   ↔   complex null line in CM   (C)

The relation (C) describes the ambitwistor transform.

Because the integrability along complex lines is less stringent than on α-planes, the ambitwistors allow to describe nonself-dual solutions, i.e. general YM fields without [6,7] and with sources [8,9], and non-self-dual metrics of general relativity [10,11]. It appears however, that the conditions characterizing the solutions of YM equations (without and with sources) are complicated, and the restriction on deformation of flat ambitwistor space which describes Einstein equations is even not known.

Both approaches described above require the complex extension of Minkowski space-time. In SUSY theories it appears however useful to consider the integrability on the SUSY generalization of

c) the space of real null lines (light-like rays) in real Minkowski space M. In particular if we put in PA z=w, we obtain the space of null twistors denoted by N, or in projective version PN. One can show that such submanifold of PA describes real null lines in M, i.e.

points in PN
                          ↔   real null lines in M   (D)
(null twistors)

Because the scalar particles in massless limit propagate along the light rays, one can also interprete <u>physically</u> null twistors as

| orbits of scalar massless particles | ⟷ | real null lines in M | (E) |

Indeed, one can show that the phase space (i.e. all observables) of free massless scalar particles can be described by a single null twistor (see e.g. [12]).

One can show that all the relations (A-E) between twistor and (complexified) space-time geometry have their supersymmetric counterparts. Supertwistors were introduced by Ferber [13], and super-ambitwistors by Witten [6]; also Witten first related the description of massless SUSY particles with supertwistors [14]. It appears that the action for massless SUSY particle contains the fermionic gauge invariance, observed firstly by Siegel [15], which describes the SUSY extension of the notion of real null (geodesic) line. The relation (E) in SUSY case is supplemented by the fermionic sector.

| fermionic gauge orbits for SUSY massless particles | ⟷ | fermionic sector of SUSY real null line | (E') |

One of the aims of extending twistor formalism to supersymmetric (SUSY) theories is a geometric explanation of superspace constraints, which are a peculiar feature of the formulation of SUSY YM and supergravity (SUGRA) theories in superspace. The integrability along real null lines does not provide any restrictions on conventional YM and gravity theories[2], but this conclusion is changed if we introduce SUSY-extended null lines, which are supplemented with additional fermionic dimensions. Using integrability along real SUSY - extended null lines there have been derived

i)   N=1,2,3   D=4   SUSY YM constraints[3] [6,16,17]

ii)   N=2   D=6 [18] and N=1   D=10 [14] SUSY YM constraints

iii) N=1   D=4 [16] and N=1   D=10 [14] SUGRA constraints

The supersymmetric extension of the integrability on α-planes (see A) was considered for the description of the super-selfduality equations and its solutions in [19-21][4]. The supersymmetric ambitwistor formalism for D=4 SUSY YM theories has been proposed in [6] and extensively studied in mathematically rigorous way by Manin [9] (see also

[23]). Recently also the constraints for D=3,4,6 and 10 SUSY YM and
SUGRA theories have been derived from the consistency of the superstring action with the D=2 generalization of Siegel fermionic invariance
(see E') [14,24,25]. Such result led to an idea (see e.g. [14,26]) that
it should be useful an introduction of the infinite-dimensional twistor space T, generalizing the relations (D,E) as follows:

orbits of massless		points of T	
strings in M	$\leftrightarrow$	(∞-dimensional)	(F)

Finally it should be mentioned that the integrability along real
null lines has been also used for the constraining of the D=2 local
string superalgebras[5].

We see that in order to relate the discussion of constraints described above with twistor formalism it is desired the following three-
fold extension of conventional twistor methods:

i)   SUSY extension
ii)   Multidimensional (Kaluza-Klein) extension to at least D=6 and
      D=10
iii) Extension to infinite-dimensional manifolds describing the
      string configurations

In this paper I shall discuss mainly the first, SUSY extension, for
D=4. In Sect.2 we shall discuss the supersymmetric extension of purely
twistor approach; in Sect.3 the supersymmetric ambitwistor space will
be introduced and the description of graded null lines. In Sect.4 we
shall only briefly discuss the multidimensional extensions of twistor
formalism for D>4. It should be mentioned that the constraints for maximally extended SUSY YM theory have neat interpretation as the integrability conditions along SUSY - extended null lines in D=10 (see also[3]). In Sect.5 we provide remarks on related problems.

2. D=4 Supertwistors.

Let us recall [1] that the basic equation expressing the incidence
between points $z_\mu \in CM$ and points in twistor space $t_A = (\omega^{\dot\alpha}, \pi_\alpha) \in T$ (μ=1,2,3,4
A=1,2,3,4; α=1,2) is

$$\omega^{\dot\alpha} = i z^{\dot\alpha\beta} \pi_\beta \qquad\qquad z^{\dot\alpha\beta} = \frac{1}{2} \sigma_\mu^{\dot\alpha\beta} z_\mu \qquad\qquad (2.1)$$

where $\sigma_\mu = (\sigma_i, 1_2)$ are Pauli matrices. The equation (2.1) describes both the correspondences (A) and (B), because

- if we hold the twistor $(\omega^{\dot\alpha}, \pi_\alpha)$ fixed one can parametrize z_μ satisfying (2.1) describing the following self-dual 2-plane

$$\alpha\text{-plane:} \qquad z^{\dot\alpha\beta} = z_0^{\dot\alpha\beta} + \lambda^{\dot\alpha}\pi^\beta \tag{2.2}$$

where $z_0^{\alpha\dot\beta}$ is any fixed point in CM satisfying (2.1), and λ^α describe two complex parameters. For any two points z_μ, $z_\mu + \Delta z_\mu$ on the same α-plane the Minkowski metric vanishes

$$\Delta z^{\alpha\dot\beta}\pi_\beta = 0 \qquad\qquad ds^2 = dz_\mu dz^\mu = 0 \tag{2.2a}$$

- if we hold $z^{\dot\alpha\beta}$ fixed and vary the twistor coordinates in (2.1), we obtain the linear 2-space in T, parametrized e.g. by Π_β. One can describe the complex 2-plane $z^{\dot\alpha\beta}$ by a pair of nonparallel twistors $t_{A;r}$ $(r=1,2)$. Introducing 2×2 matrices

$$\Omega = \begin{pmatrix} \omega^{1;}{}_1 & , & \omega^{1;}{}_2 \\ \omega^{2;}{}_1 & , & \omega^{2;}{}_2 \end{pmatrix} \qquad\qquad \Pi = \begin{pmatrix} \pi_{1;1} & \pi_{1;2} \\ \pi_{2;2} & \pi_{2;2} \end{pmatrix} \tag{2.3}$$

one can express the matrix $Z = \{z^{\dot\alpha\beta}\}$ in terms of two twistor coordinates as follows

$$Z = \Omega\Pi^{-1} \tag{2.4}$$

The formula (2.4) describes composite structure of CM in terms of the coordinates on bitwistor space $T \times T^{6)}$.

In twistor space T one can define the action of SL(4;C) (4-fold cover of the conformal group of CM), or in particular SU(2,2) (4-fold cover of the conformal group of M), and introduce the U(2,2)-invariant scalar product

$$\langle u, t \rangle = \bar{u}_A G_{AB} t_B \tag{2.5}$$

where G_{AB} is a Hermitean metric with the signature (++--). Choosing

$$G_{AB} = \begin{pmatrix} 0 & I_2 \\ I_2 & 0 \end{pmatrix} \tag{2.6}$$

and putting $u = (\eta^\alpha, \rho_\alpha)$ one obtains

$$\langle u,t \rangle = \bar{\rho}_{\dot{\alpha}}\omega^{\dot{\alpha}} + \bar{\xi}^{\alpha}\pi_{\alpha} \tag{2.7}$$

The form (2.7) of the scalar product (2.5) exhibits the decomposition of D=4 twistor as SO(4,2) spinor into Weyl spinor and Weyl cospinor of Lorentz group O(3,1). The interpretation of twistor as a D=4 conformal spinor permits to express the manifold of complex 2-planes as the following Hermitean coset space

$$CM \simeq G_2(C^4) = \frac{SU(2,2)}{S(U(2) \times U(2))} \tag{2.8}$$

The formula (2.8) permits to derive the conformal transformations of $Z=z_n\sigma_n$ as 2×2 matrix Möbius transformation (see e.g. [31])

$$Z = \frac{A + BZ}{C + DZ} \qquad \begin{pmatrix} A & B \\ C & D \end{pmatrix} \in SU(2,2) \tag{2.9}$$

confirming the interpretation of z_μ as the coordinates of (compacti-fied) CM.

The twistors can be defined in several ways, equivalent for D=4, e.g. as

i)   fundamental D=4 conformal spinors

ii) the solution space of the "twistor equation", defining confor-mal Killing spinors (such a definition is related closely with formula (2.1).

iii) four complex coordinates describing the phase space of free massless conformal particles (see e.g. [12,32,33]).

v)   twistor bundle over space-time M with the fibre described by all complex structures on M (see e.g. [34]).

Following Ferber [13] we shall extend here supersymmetrically the first definition. The N-extended conformal superalgebra SU(2,2;N) is obtained by adding to SU(2,2) generators 4N complex supercharges and N^{2} [7] internal U(N) generators (see e.g. [34,35]). The fundamental representation of SU(2,2;N) is described by (4+N)-dimensional complex superspace of supertwistors $(t_A, \xi_i) \in T_{(N)} = C^{4;N}$ (i=1,... N), with 4 even (commuting) and N odd (anticommuting) coordinates. The U(2,2) norm (2.5) is extended as follows:

$$\langle u_{(n)}, t_{(N)} \rangle = \bar{u}_A G_{AB} t_B + \bar{\eta}_i \xi_i \tag{2.10}$$

where $U_{(N)} = (u_A, \eta_i)$. The superalgebra of SU(2,2;N) is realized on $T_{(N)}$

by $(4+N) \times (4+N)$ matrices [13,35], and $SU(2,2;N)$ matrix supergroup is obtained by the exponentiation map with commuting parameters in the bosonic sector, and anticommuting in the fermionic one (see e.g. [36,37]).

In supertwistor space the correspondences (A) and (B) becomes non-
unique because one can introduce N+1 superspaces by the following SUSY
extension of the formula (2.8) [38-40]

$$SCM_k^{(N)} \simeq G_{2;k}(C^{4;N}) = \frac{SU(2,2;N)}{S(U(2,k) \times U(2,N-k))} \qquad (2.11)$$

where $SCM_0^{(0)} \equiv CM$, and $k=0,1...N$.
The SUSY version of the relation (B) can be written separately for every N-extended superspace $SCM_k^{(N)}$ as follows

$$\begin{array}{l} \text{linear } (2;k) \text{ subspaces in } T_{(N)} \\ (\text{linear } (1;k) \text{ subspaces in } PT_{(N)}) \end{array} \leftrightarrow \text{points in } SCM_k^{(N)} \qquad (2.12)$$

where linear subspaces (n,m) are parametrized by n even and m odd coordinates. The basic formula (2.1) is extended for the superspace $CM_k^{(N)}$
in the following way [40]

$$\omega^{\dot{\alpha}} = iz^{\dot{\alpha}\beta}\pi_\beta + \sum_{j=1}^{k} \Theta^{\dot{\alpha}j}\xi_j$$

$$\xi_1 = \Theta^\alpha_1 \pi_\alpha + \sum_{j=1}^{k} \lambda_1{}^j \xi_j \qquad (2.13)$$

where $l=k+1,...N$, and the coordinates of $CM_k^{(N)}$ are

$$SCM_k^{(N)}: \quad (z_\mu, \lambda_1{}^j; \Theta^\alpha_1, \Theta^{\dot{\alpha}j}) \qquad (2.14)$$

i.e. $4+k(N-k)$ even and 2N odd coordinates. We see that only for k=0 and
k=N the even sector is given by CM. In such a case the formulae (2.13)
are simplified, and the equations for super-α-plane are[8]

$$\text{chiral superspace} \qquad \omega^{\dot{\alpha}} = iz_+^{\dot{\alpha}\beta}\pi_\beta \qquad \xi_i = \Theta^\alpha_i \pi_\alpha \qquad (2.15a)$$
$$(k=0)$$

$$\text{antichiral superspace} \qquad \omega^{\dot{\alpha}} = iz_-^{\dot{\alpha}\beta}\pi_\beta + \Theta^{\dot{\alpha}i}\xi_i \qquad (2.15b)$$
$$(k=N)$$

where we introduced different chiral and anti-chiral complex Minkowski coordinates, because in the formulae (2.14) a priori the coordinates of $CM_k^{(N)}$ for different values of k are not related.

The formulae (2.13) for fixed values of the supertwistor coordinates $(\omega^\alpha, \pi_\alpha, \xi_i)$ describe the super-α-plane (2,k) in $CM_k^{(N)}$. For k=0 and k=N one obtains the following parametric equations for super-α-planes:

chiral (k=0)
super-α-plane
(2,N)

$$z_+^{\dot\alpha\beta} = z_{(0)+}^{\dot\alpha\beta} + \lambda^{\dot\alpha}\pi^\beta$$

$$\theta_i^\alpha = \theta_{1(0)}^\alpha + \varepsilon_i\pi^\alpha$$

(2.16a)

antichiral (k=N)
super-α-plane
(2,2N)

$$z_-^{\dot\alpha\beta} = z_{(0)-}^{\dot\alpha\beta} + c\lambda^{\dot\alpha}\pi_\beta + i\varepsilon_i^{\dot\alpha}\theta_1^\beta$$

$$\theta^{\dot\alpha i} = \theta_{(0)}^{\dot\alpha i} + \varepsilon^{\dot\alpha i}$$

(2.16b)

where the coordinates with subscript "0" denote particular solutions and λ^α are complex, ε_i, $\varepsilon_i^{\dot\alpha}$ complex-Grassmann parameters. It should be stressed that θ_i^β occuring in (2.16b) is a <u>chiral</u> coordinate, defined by (2.15a)[9]. It makes therefore sense to consider nonchiral superspace, described by the coordinates $(z_+, z_-, \theta_i^\alpha, \theta_i^{\dot\alpha})$. It appears from (2.15a,b) that

$$\omega^{\dot\alpha} = iz_+^{\dot\alpha\beta}\pi_\beta = i(z_-^{\dot\alpha\beta} - i\theta^{\dot\alpha i}\theta_i^\beta)\pi_\beta$$

(2.17)

i.e. one can identify[10]

$$z_+^{\dot\alpha\beta} = z_-^{\dot\alpha\beta} - i\theta^{\dot\alpha i}\theta_i^\beta$$

(2.18)

or define the "symmetric" CM coordinate $z^{\dot\alpha\beta}$ by[11]

$$z_\pm^{\dot\alpha\beta} = z^{\alpha\beta} \mp \frac{i}{2}\theta^{\alpha i}\theta_i^\beta$$

(2.19)

The complexified nonchiral superspace $SCM^{(N)}$ is described by the superspace coordinates $(z^{\dot\alpha\beta}, \theta^{\dot\alpha i}, \theta_i^\alpha)$. The α-plane in $SCM^{(N)}$ is determined by the eq.(2.15a). Using symmetric CM coordinates (see (2.19)) we obtain

nonchiral
super-α-plane
(2,3N)

$$z^{\dot\alpha\beta} = z_{(0)}^{\dot\alpha\beta} + \lambda^{\dot\alpha}\pi^\beta + \frac{i}{2}(\delta\theta^{\dot\alpha i}\theta_i^\beta - \theta^{\dot\alpha i}\delta\theta_i^\beta)$$

$$\delta\theta_i^\alpha = \theta_i^\alpha - \theta_{i(0)}^\alpha = \varepsilon_i\pi^\alpha$$

$$\delta\theta^{\dot\alpha i} = \theta^{\dot\alpha i} - \theta_{(0)}^{\dot\alpha i} = \varepsilon^{\dot\alpha i}$$

(2.20)

where $\lambda^{\dot\alpha}, \varepsilon_i$ and $\varepsilon^{\dot\alpha i}$ are the parameters.

The integrability of SUSY YM superspace connection forms on nonchiral super-α-planes (2.20) was used for the SUSY generalization of the self-duality equation (see [21]).

3. D=4 SUSY ambitwistors

In order to describe geometrically an ambitwistor we introduce for a dual twistor $\bar{u}_A = (\rho^\alpha, r_{\dot\alpha})$ (where $\rho^\alpha = (\rho^{\dot\alpha})^*$, $r_{\dot\alpha} = r_\alpha^*$) the dual incidence equation

$$\rho^\beta = - i r_{\dot\alpha} z^{\dot\alpha\beta} \qquad (3.1)$$

which determines in CM the anti-self-dual β-plane, parametrized as follows

$$\beta - \text{plane} : \qquad z^{\dot\alpha\beta} = z_0^{\dot\alpha\beta} + r^{\dot\alpha} \lambda^\beta \qquad (3.2)$$

where λ^β are two complex parameters.

Let us assume now that the complex point $z^{\dot\alpha\beta}$ lies simultaneously on α-plane described by twistor t_A (see eq.(2.1)) and α-plane described by dual twistor $\bar{u}_A$ (see eq.(3.1)). Multiplying (2.1) by $\rho_{\dot\alpha}$, and (3.1) by π_β one obtains the consistency condition

$$r_{\dot\alpha} \omega^{\dot\alpha} + \rho^\alpha \pi_\alpha = <u,t> = 0 \qquad (3.3)$$

The solution of the eq.(2,1) and (3.1) with (3.3) provides the parametric equation for the complex null line

$$z^{\dot\alpha\beta} = i \frac{\omega^{\dot\alpha} \rho^\beta}{r_{\dot\alpha} \omega^{\dot\alpha}} + c r^{\dot\alpha} \pi^\beta \qquad (\text{c complex}) \qquad (3.4)$$

i.e. we obtain the correspondence (C). If the coordinates z_μ are real, the eq.(2.1) and (3.1) have common real points in M only if $t_A = u_A$, and the eq.(3.4) takes the form[12]

$$x^{\dot\alpha\beta} = i \frac{\omega^{\dot\alpha} \bar{\omega}^\beta}{\omega^{\dot\alpha} \bar{\pi}_{\dot\alpha}} + \lambda \bar{\pi}^{\dot\alpha} \pi^\alpha \qquad (\lambda \text{ real}) \qquad (3.5)$$

and we arrive at the correspondence (D).

The SUSY extension of ambitwistor (1.1) is given by [6,9,41]

D=4 complex superambitwistor $PA_{(N)} = (U_{(N)}, t_{(N)}) \subset PT_{(N)} \times \overline{PT}_{(N)}$
      space:

$$\langle U_{(N)}, t_{(N)} \rangle_C = 0 \qquad (3.6)$$

where the scalar product is given by (2.10), and projective twistors
are defined modulo complex scale factor. Explicitly we have

$$r_{\dot\alpha}\omega^{\dot\alpha} + \rho^\alpha \pi_\alpha + \eta^i \xi_i = 0 \qquad (3.7)$$

where $\lambda^i = \bar\lambda_i$. The SUSY extension of the eq.(3.1) looks as follows

$$\rho^\beta = -ir_{\dot\alpha}z^{\dot\alpha\beta} + \sum_{l=k+1}^{N} \Theta_l^\beta \eta^l \qquad (3.8)$$

$$\eta^j = \Theta^{\dot\alpha j}r_{\dot\alpha} + \sum_{l=k+1}^{N} \lambda_l^j \eta^l$$

We see from (2.13) and (3.8) that in every superspace $CM_k^{(N)}$ one can introduce (2;N-k)-dimensional super-β-planes and (2,k) dimensional super-
β-planes. In particular for k=0 and k=N one obtains the following equations for super-β-planes:

chiral superspace $\qquad \rho^\beta = -ir_{\dot\alpha}z_+^{\dot\alpha\beta} + \Theta_i^\beta \eta^i \qquad (3.9a)$
(k=0)

antichiral superspace: $\qquad \rho^\beta = -ir_{\dot\alpha}z_-^{\dot\alpha\beta} \qquad \eta^i = \Theta^{\dot\alpha i}r_{\dot\alpha} \qquad (3.9b)$
(k=N)

which can be parametrized as follows

chiral (k=0) $\qquad z_+^{\dot\alpha\beta} = z_{+(0)}^{\dot\alpha\beta} + r^{\dot\alpha}\lambda^\beta - i\Theta^{\dot\alpha i}\varepsilon_1^\beta$
super-β-plane $\qquad\qquad\qquad\qquad\qquad\qquad\qquad\qquad\qquad (3.10a)$
(2,2N) $\qquad\qquad \Theta_i^\alpha = \Theta_{i(0)}^\alpha + \varepsilon_i^\alpha$

antichiral (k=N) $\quad z_-^{\dot\alpha\beta} = z_{-(0)}^{\alpha\beta} + cr^{\dot\alpha}\lambda^\beta$
super-β-plane $\qquad\qquad\qquad\qquad\qquad\qquad\qquad\qquad\qquad (3.10b)$
(2,N) $\qquad\qquad \Theta_1^{\dot\alpha i} = \Theta_{(0)}^{\dot\alpha i} + \varepsilon^i r^{\dot\alpha}$

where $\lambda^\beta, \varepsilon^i$ and ε_i^β are the parameters.
The β-plane in complexified nonchiral space is defined by the eq.(3.9b)
with additional requirement (2.19). We obtain

nonchiral super-β-plane $\quad z^{\dot\alpha\beta} = z_{(0)}^{\dot\alpha\beta} + r^{\dot\alpha}\lambda^\beta + \frac{1}{2}(\delta\Theta^{\dot\alpha i}\Theta_1^\beta - \Theta^{\dot\alpha i}\delta\Theta_1^\beta)$
(2,3N)

$$\delta\Theta_1^\alpha = \Theta_1^\alpha - \Theta_{1(0)}^\alpha = \varepsilon_i^\alpha$$

$$\delta\Theta^{\dot\alpha i} = \Theta^{\dot\alpha i} - \Theta_{(0)}^{\dot\alpha i} = \varepsilon^i r^{\dot\alpha} \tag{3.11}$$

It can be checked that the point $(z^{\alpha\dot\beta}, \Theta_i^\alpha, \Theta_i^{\dot\alpha})$ can lie on super-α-plane (2.20) and super-β-plane (3.11) simultaneously if the condition (3.7) is valid. The superambitwistor (3.6) describes the following SUSY extension of complex null line (3.4), with the complex dimension (1,2N)

$$z^{\dot\alpha\beta} = z_{(0)}^{\dot\alpha\beta} + cr^{\dot\alpha}\pi^\beta + \frac{i}{2}(\delta\Theta^{\dot\alpha i}\Theta_i^\beta - \Theta^{\dot\alpha i}\delta\Theta_i^\beta) \tag{3.12}$$

$$\delta\Theta_i^\alpha = \varepsilon_i\pi^\alpha \qquad \delta\Theta^{\dot\alpha i} = \varepsilon^i r^{\dot\alpha}$$

where c, ε_i and ε^i are complex parameters.

It is easy to see that all the translations along super-α-plane (2.20), super-β-plane (3.11) and consequently along (3.12) are SUSY-null, i.e. they lead to vanishing of the following SUSY-extended Minkowski metric (compare with 2.2a)[13]

$$\omega^{\dot\alpha\beta}\omega_{\dot\alpha\beta} = 0 \qquad \omega^{\alpha\beta} = dz^{\dot\alpha\beta} - \frac{i}{2}(d\Theta^{\dot\alpha i}\Theta_i^\beta - \Theta^{\dot\alpha i}d\Theta_i^\beta) \tag{3.13}$$

Finally it follows from the eq.(2.15a) and (3.9b) that the points of real chiral superspace SM=$(x_\mu, \Theta_i^\alpha, \Theta^{\dot\alpha i} = (\Theta_i^\alpha)^*)$ lie simultaneously on super-α-plane and super-β-plane if $U_{(N)} = t_{(N)}$, i.e. when SUSY ambitwistor "collapses" to a null supertwistor. The null supertwistors describe real SUSY-null line, extending by 2N real Grassmann dimensions the formula (3.5) as follows:

$$x^{\dot\alpha\beta} = x_{(0)}^{\dot\alpha\beta} + \lambda\bar\pi^{\dot\alpha}\pi^\alpha + \frac{i}{2}(\delta\Theta^{\dot\alpha i}(\Theta_i^\beta)^* - \Theta^{\dot\alpha i}(\delta\Theta_i^\beta)^*) \tag{3.14}$$

$$\delta\Theta_i^\beta = \varepsilon_i\pi^\beta \qquad \delta\Theta^{\dot\beta i} = (\delta\Theta_i^\beta)^*$$

It was observed firstly by Witten [6] and further investigated in detail by Manin [9] that for N-extended SUSY YM system ($0 \leq N \leq 3$) the solutions of SUSY YM equations can be obtained as permitting the extensions to (3-N)-th order of the infinitesimal neighbourhood in the space of SUSY-extended null lines (3.14). In particular if N=3 the integrability on (3.14) is equivalent to the equations of motion. In the case

N=4 the self-duality condition in internal O(4) space

$$\phi^{ij} = \frac{1}{2} \varepsilon^{ijkl} \phi_{kl} \tag{3.15}$$

obstructed the twistorial interpretation in D=4. However, this difficulty disappears if we consider N=1 D=10 SUSY YM theory [14] which provides N=4 D=4 SUSY YM equations via dimensional reduction. In such a
way it became interesting to consider the SUSY twistor formalism in D>4.
Another reason for considering D>4 is a possible relation between twistor formalism and string theories [41-44].

4. D > 4

The D=4 twistors can be defined in several equivalent ways, for
example

  i)   as the fundamental representation of the four-fold covering
SU(2,2)=$\overline{SO(4,2)}$ of D=4 conformal group.

  ii) as the parametrization of the totally null 2-planes in complexified Minkowski space C^4

  iii) as a bundle over S^4 describing all possible complex structures on S^4

We shall discuss briefly the extension of these definitions to D>4.

  i) Twistors as conformal spinors.

We define twistors as fundamental spinors of SO(D,2).

In such a way one can introduce twistors for any D, and (depending
on the choice of D) they can be real, complex or quaternionic. They are
described by the following linear vector spaces for 4≤D≤10 (see e.g.
[45])

D	4	5	6	7	8	9	10
T	C^4	H^4	H^4	H^8	C^{16}	R^{32}	R^{32}

Table 1. D-dimensional twistors as the fundamental conformal
        spinors

It should be added that for any D the conformal spinors are described by a pair of Lorentz spinors. This decomposition is valid for complexified rotation groups as well as for the real one, with arbitrary
signature.

In such a framework the dimensions D=6 and D=10 are selected because they correspond to quaternionic and octonionic extensions of the complex descriptions of D=4 spinors and D=4 twistors. Let us write the following table (see e.g. [46])

spin covering of Lorentz group	D=4 $SL(2,C)$	D=6 $SL(2;H)$	D=10 $DL(2;0)$
Weyl spinors	C^2	H^2	$R^{16} \simeq 0^2$
spin covering of Conf.group	$SU(2,2) = =U_\alpha(4;C)$	$U_\alpha(4;H)$	$U_\alpha(4;0)$
Conf.fundamental spinors	C^4	H^4	$R^{32} \simeq 0^4$

Table 2. The relation of complex numbers with D=4, quaternions with D=6 and octonions with D=10

where (F=C,H or O)

$$U_\alpha(4;F) : \quad \bar{q}_A H_{AB} q_B = inv \qquad H^+ = -H$$

i.e. U_α describes antiunitary group. In particular one can chose $H=\begin{pmatrix} 0 & 1 \\ -1 & 0 \end{pmatrix}$

ii) Twistors as pure spinors.

Following [1,42-44] one can adopt the view that in even dimensions D=2k twistors are pure conformal spinors, describing totally null k-planes in C^{2k}. We obtain the following generalization of the correspondence (A) for even D>4:

$$\text{point in } T_{(2k)} \quad \leftrightarrow \quad \text{totally null k-planes in } C^{2k} \qquad (A')$$

where $T_{(2k)}$ denotes the space of twistors for D=2k.

Pure spinors are obtained by imposing $r=2^k - \frac{k(k+1)}{2}$ linear constraints on "ordinary" projective fundamental SO(D+2;C) spinors, with complex dimension $n=2^k-1$ i.e. they are described by quadric Q^q, with complex dimension $q= \frac{k(k+1)}{2}$. We obtain the following complex manifolds describing "ordinary" projective spinors and twistors:

	"ordinary" projective conf. spinors	twistors
D=4	CP(3)	CP(3)
D=6	CP(7)	Q^6

	"ordinary" projective conf. spinors	twistors
D=8	CP(15)	Q^{10}
D=10	CP(31)	Q^{15}

Table 3. From ordinary to pure conformal spinors.

We see that for D=4 ordinary and pure twistors can be identified, for D=6 one needs to impose one constraint, for D=8 r=5, and for D=10 r=16. In particular for D=6 it can be shown [42] that the purity condition follows as the consistency condition for the Penrose incident equation (2.1) extended from D=4 to D=6.

iii) <u>Twistor space as bundle over S^{2k} describing all possible complex structures.</u>

In the description of self-dual fields on (compactified) Euclidean space one can extend the gauge connections from S^4 to CP(3) provided the extension is pure gauge. Such a construction leads to the introduction of twistor space as fiber bundle over S^4 with fibers described locally by CP(1)= $\frac{SU(2)}{U(1)}$. Because $\overline{SO(4)}$=SU(2)×SU(2), one can write also that CP(1)= $\frac{\overline{SO(4)}}{U(2)}$, i.e. locally PT $\simeq S^4 \times \frac{SO(4)}{U(2)}$. This relation can be extended to any even k and we obtain locally for D=2k (see e.g. [47-49])

$$PT \simeq S^{2k} \times \frac{SO(2k)}{U(k)} \qquad (4.1)$$

Counting real dimensions $2k+k(2k-1)-k^2=k(k+1)$, we get the agreement with the dimensions obtained from our second definition (i.e. twistors as pure spinors). From the identification of the definitions ii) and iii) one gets

D=4	D=6	D=8	D=10
DP(3)	Q^6	Q^{10}	Q^{15}
↓CP(1)	↓CP(3)	↓ Q^6	↓ Q^{10}
S^4	S^6	S^8	S^{10}

Table 4. Twistor bundles in D=2k (k=2,3,4,5).

The twistor bundles written above should be useful for the description of selfdual gauge fields in D=2k, k=2,3,4,5.

Finally we shall consider

iv) Supersymmetrization of twistors for D>4

Only for D=6 the spin coverings $\overline{SO(D,2)}$ (D>4) is described by a classical Lie group, and $\overline{SO(6,2)} = U_\alpha(4;\overset{\ast}{H})$ [50]. The D=6 conformal spin group can be supersymmetrized as follows [51]

$$U_\alpha(4;H) \rightarrow U_\alpha U(4;H) \tag{4.2}$$

where the bosonic sector of the SUSY extension of D=6 conformal group is

$$U_\alpha(4;H) \times U(N;H) = SO^*(8) \times Sp(2n) \tag{4.3}$$

The D=6 conformal superspinors are described by the SUSY extension of H^4

$$H^4 = (q_1 \ldots q_4) \rightarrow H^{4;N} = (q_1 \ldots q_4; \Theta_1 \ldots \Theta_N) \tag{4.4}$$

where $\Theta_i = \Theta_i^o + \Theta_i^r e_r$ are Grassmann-valued quaternionic coordinates[14].

We see therefore that in the framework of conventional Z_2-graded superalgebras [53] one can only supersymmetrize the twistors as "ordinary" conformal spinors for D=6. It is not clear how the condition of purity can be supersymmetrized. The supersymmetrization of D=10 conformal algebra described by 4×4 antihermitean matrices [46] leads beyond the framework of associative Lie superalgebras (see [54]).

5. Final Remarks.

We would like to make the following comments:

i) There exists a close relation between twistor methods and harmonic approach to extended SUSY [55,56]. The additional bosonic scalar variables in N=2,3 harmonic superspace lie on the quadric $Q_{2N-3}=$ $((u_i,v_i) \in CP^{N-1} \times CP^{N-1}, u_i v^i=0)$, which recalls the ambitwistor space[15]. The discussion of harmonic superspace exploiting some notions of twistorial point of view has been presented in [49,57].

ii) Massless strings propagate along null lines, and their dynamics is described by actions selecting minimal world sheets as "string trajectories". In particular one can show that in any dimension the propagation of a bosonic string is described by a pair of null curves [58, 43]. The twistorial parametrization of null curves has been given for D=3 [59] and D=4 [58], and the parametric formulae of Weierstrass [59] for D=3 and Montcheuil [60], Eisenhart [61] for D=4 has been derived.

The explicite parametrization of null curves for D=6 has been considered by Hughston and Shaw [42], and the application of supertwistors to the description of superstrings in D=4 complex nonchiral superspace has been given in [41].

iii) The "strong version" of twistor approach to physics is the replacement of QFT in Minkowski space by QFT in twistor or ambitwistor space (see e.g. [62, 63]). Such a programm was also investigated in the framework of Rzewuski's spinor space approach [64,65]. It should be also observed that there is an analogy between

- composite nature of space-time coordinates, which can be localized only in terms of two twistors (see (2.4))

- composite nature of elementary objects with unobservable constituents (quark or preon models).

It is tempting to describe the confinement of unobservable constituents as due to the fact that they are in some sense localized in twistor space. Some investigations related with such an idea were made in [66-68], where the notion of quark-twistor variables was proposed, and they were related with strings on conformal supergroup manifold.

Acknowledgments

The author would like to thank dr. L. Hughston for several valuable discussions, and Prof. A. Salam for the hospitality at the International Centre for Theoretical Physics, where these lecture notes were completed. We would like also to mention that the first version of this talk was presented at the First Torino Meeting on Unification and Superunification (23.IX - 27.IX 1985).

This paper I would like to dedicate to my teacher, prof. J. Rzewuski on his 70^{-th} birthday, who taught me first that the spinor coordinates should be more fundamental than the ones described by space-time fourvectors.

FOOTNOTES

1. These generalizations were defining "googly" states (photons, gravitons); see e.g. [4,5].
2. One can show that the coupling of massless particles to external YM and gravity fields does not impose any restrictions on these background fields.
3. For N=4 D=4 SUSY YM theory it is not known how to derive from the integrability along SUSY-extended null lines the internal sector selfdu-

ality constraint for the field strenght superfield.

4. It should be mentioned however, that in these papers the structure of SUSY in real Euclidean space has not been taken into account. For the discussion of super-self-duality with more explicite discussion of Euclidean SUSY see [22].

5. For local string superalgebra see [27] , for integrability see [28].

6. This formula is due to Penrose, but some authors did put forward earlier ideas that space-time coordinate can be expressed as composite in terms of spinor components - see e.g. [29] (see also [30]).

7. For N=4 one gets as internal symmetry groups SU(4) [34].

8. We call superspace $SCM_0^{(N)}$ chiral because the variables θ_i^α and $\theta_i^{\dot\alpha}$ can be obtained from 4-component complex Dirac spinor by chiral projections $\frac{1}{2}(1\pm\gamma_5)$ (for γ_5 diagonal).

9. From (2.15b) one gets for the last term of $\delta z^{\dot\alpha\beta}$ the equation $i\delta z^{\dot\alpha\beta}\pi_\beta+ +\delta\theta_i^{\dot\alpha}\xi_i=0$ which is solved by $\delta z^{\dot\alpha\beta}=i\delta\theta_i^{\dot\alpha}\theta_i^\beta$ if we put $\xi_i=\theta_i^\alpha\pi_\beta$.

10. The formula (2.18) can be explained geometrically as supersymmetric flag manifold (see e.g. [9]).

11. In formula (2.19) one can recognize the known relation between the real, chiral and antichiral superspace coordinates

12. From (3.3) follows that $\omega^{\dot\alpha}\bar\pi_\alpha$ is purely imaginary.

13. The SUSY-extended null lines defined in [6] in fact do not lie on the super-light-cone (3.13), because the part describing the translations along Grassmann directions is missing. This simplification does not invalidate however the conclusions in [6].

14. The quaternionic supergroups as quaternionic norm-preserving endomorphisms in superspace were considered recently in [52].

15. In ref. [57] even the name "isotwistor superspace was proposed as more appropriate than "harmonic superspace".

REFERENCES

1. R. Penrose and W. Rindler, "Spinors and Space-Time", Vol.2,Cambridge Univ.Press, 1986, and the literature quoted therein
2. R. Ward, Phys.Lett.A61,81,1977
3. M.F. Atiyah, V.G. Drinfeld, N.I.Hitchin and Yu.I.Manin, Phys.Lett. A65,185(1978)
4. R. Penrose in "Advances in Twistor Theory", ed.L.P.Hugston and R.S. Ward, Pitman,London, 1979
5. L.J. Mason, Twistor Newsletters,No.19 and 20
6. E. Witten, Phys.Lett.77B,394(1978)
7. J.Isenberg, P.B.Yasskin and P.Green, Phys.Lett.78B,462(1978)
8. G.M. Henkin and Yu.I.Manin, Phys.Lett.95B,405(1980)
9. Yu.I.Manin "Gauge fields and complex geometry", ed.Nauka, Moscow 1984 (in Russian)
10. J. Isenberg and P. Yasskin, Gen.Rel.Grav.14,621(1982)

11. C.R. Le Brun, Class.Quantum Grav.$\underline{2}$,555(1985)
12. L.P. Hughston, "Twistors and Particles", Lect.Notes in Phys.No 79 (Springer),1979
13. A. Ferber, Nucl.Phys.$\underline{B132}$,55(1978)
14. E. Witten, Nucl.Phys.$\underline{B266}$,245(1986)
15. W. Siegel, Phys.Lett.$\underline{128B}$,397(1983)
16. J. Crispin-Romão, A. Ferber and P. Freund, Nucl.Phys.$\underline{B182}$,45(1981)
17. I.V. Volovich, Teor.Math.Fiz.$\underline{54}$,89(1983) (in Russian)
18. C. Devchand, "Integrability on light-like lines in six-dimensional superspace", Freiburg Univ.preprint,1986
19. Yu.I. Manin in "Problems of High Energy Physics and QFT, Proc.Protvino Seminar 1982,p.46
20. A.M. Semikhatov, Phys.Lett.$\underline{120B}$,171(1983)
21. I.V. Volovich, Teor.Math.Fiz.$\underline{55}$,39(1983) (in Russian)
22. J. Lukierski and W. Zakrzewski, to appear as ICTP preprint
23. A.A. Rosly, Class.Quantum Grav.$\underline{2}$,693(1985)
24. M.T. Grisaru, P.S. Hove, L. Mezincescu, B.E.W. Nillson and P.K. Townsend, Phys.Lett.$\underline{162B}$,116(1985)
25. E. Bergshoff, E. Sezgin and P.K. Townsend, Phys.Lett.$\underline{169B}$,191(1986)
26. J. Isenberg and P. Yasskin, Ambitwistors (and strings?), preprint 1986
27. W. Siegel, Nucl.Phys.$\underline{B263}$,93(1985)
28. P.G.O. Freund and L. Mezincescu, preprint EFI 86-11(1986)
29. J. Rzewuski, Nuovo Cim.$\underline{5}$,942(1958)
30. J. Kocik and J. Rzewuski, "On projections of spinor spaces onto Minkowski space", to be published in "Symmetries in Science II", ed. B. Gruber, Plenum Press, New York, 1986
31. R.O. Wells Jr. Bull.Am.Math.Soc.(New Serie) $\underline{1}$,296(1979)
32. W. Lisiecki and A. Odzijewicz, Lett.Math.Phys.$\underline{3}$,325(1979)
33. I.T. Todorov, "Conformal description of Spinning particles", SISSA preprint 1/81
34. R. Haag, J. Łopuszański, and M. Sohnius, $\underline{B88}$,257(1975)
35. S. Ferrara, M. Kaku, P. van Nieuvenhuizen and P.K.Townsend, Nucl. Phys.$\underline{B129}$,125(1977)
36. F.A. Berezin, ITEP preprint ITEP-76,1977
37. F. Gursey and L. Marchildon, J.Math.Phys.$\underline{19}$,942(1979)
38. J. Lukierski, "From supertwistors to composite superspace", Wrocław Univ.preprint 534, 1981
39. L.B. Litov and V.N. Pervushin, Phys.Lett.$\underline{B147}$,76(1984)
40. M. Kotrla and J. Niederle, Czech.J.Phys.$\underline{B35}$,602(1985)
41. W.T. Shaw, Class.Quantum Grav.$\underline{3}$,753(1986)
42. L.P. Hughston and W.T. Shaw, "Minimal Curves in Six Dimensions",MIT preprint, 1986
43. P. Budinich, "Null vectors, spinors and strings", SISSA preprint 10/86
44. W.T. Shaw, "Classical Strings and Twistor Theory: How to solve string equations without using the light-cone gauge", Talk at VC Santa Cruz AMS meeting, June 1986
45. T. Kugo and P. Townsend, Nucl.Phys.$\underline{B221}$,357(1983)
46. A. Sudbury, J.Phys.$\underline{A17}$,939(1984)
47. R.L. Bryant, Duke Math.J. $\underline{52}$,223(1985)
48. A.M. Semikhatov, JETP Letters $\underline{41}$,201(1985)
49. A.M. Semikhatov, "Harmonic superspaces and the division algebras", Lebedev Inst.preprint N^O 339(1985)
50. R. Gilmore, "Lie groups, Lie algebras and some of their applications", Willey, New York, 1984
51. Z. Hasiewicz, J. Lukierski and P. Morawiec, Phys.Lett.$\underline{130B}$,55(1983)
52. J. Lukierski and A. Nowicki, Ann.of Phys.$\underline{166}$,164(1986)
53. V. Kac, Comm.Math.Phys.$\underline{53}$,31(1977)
54. Z. Hasiewicz and J. Lukierski, Phys.Lett.$\underline{145B}$,65(1984)
55. V.I. Ogievetski and E.S. Sokhaczew, Yadernaja Fiz.$\underline{31}$,205(1980)
56. A. Galperin, E.Ivanov, S.Kalitzin, V.O.Ogievetski and E.Sokhaczew

Class.Quantum Grav.$\underline{1}$,469(1984)
57. A.A.Rosly and A.S.Schwarz, "Supersymmetry in a space with auxiliary dimensions", ITEP preprint 39/1985
58. W.T. Shaw, Class.Quantum Grav.$\underline{2}$,L113(1985)
59. K. Weierstrass, Monats.Berl.Acad.$\underline{612}$(1866)
60. M. Montcheuil, Bull.Soc.Math.France $\underline{33}$,170(1905)
61. L.P. Eisenhart, Ann.Math.(Ser.II),$\underline{13}$,$\underline{17}$(1911)
62. M.A.H. Mac Callum and R. Penrose, $\underline{Phys}$.Rep.$\underline{6}$,241(1972)
63. A.P. Hodges, Proc.R.Soc.Lond.$\underline{A397}$,375(1985)
64. J. Rzewuski, Acta Phys.Polon.$\underline{18}$,549(1959)
65. J. Rzewuski, Rep.Math.Phys.$\underline{22}$,235(1985)
66. J. Lukierski, Lett.Nuovo Cim.$\underline{24}$,309(1979)
67. J. Lukierski in "Hadronic Matter at Extreme Energy Density", ed.by
N. Cabbibo and L. Sartorio, Plenum Press, 1980,p.187
68. J. Lukierski, J.Math.Phys.$\underline{21}$,561(1980)

SUPERSYMMETRIES OF THE DYON [*]

Eric D'Hoker
Department of Physics
Princeton University
Princeton, New Jersey 08544
U.S.A.

Luc Vinet
Laboratoire de Physique Nucléaire
Université de Montréal
C.P. 6128 Succ. "A"
Montréal, Québec
H3C 3J7 Canada

Contents

---

[*] Seminar delivered by Luc Vinet in January 1986 at the Laboratoire de Physique Théorique et des Hautes Énergies, Université Pierre et Marie Curie (Paris VI).

INTRODUCTION

Over the last two years or so, we have investigated the rôle of superalgebras as dynamical algebras in Quantum Mechanics[1]. The first problem we analyzed[2,3] was that of a Non-Relativistic spin-1/2 particle in the field of a Dirac magnetic monopole which was shown to possess an OSp(1,1) dynamical superalgebra. We also observed[4] that this system can be generalized to accomodate a $1/r^2$-potential and further noted the presence of an N=2 superconformal symmetry in such instances. These interesting observations allowed us to obtain the spectrum and wave functions of the above systems from group theory alone.

A famous problem with dynamical symmetries is certainly that of a spinless charged particle in a Coulomb potential. It possesses an O(4) invariance algebra which explains the "accidental" degeneracy of the spectrum and all its states fall into a single irreducible representation of O(4,2). A natural question that one can ask then, is the following: Can we find supersymmetries in the presence of a 1/r-potential? We came up with the following answer.

A. Consider the Hamiltonian

$$H_D = \frac{1}{2}\left[P_i - (q - \frac{1}{2}\Sigma)A_i^D\right]^2 - \frac{1}{r}$$

$$+ \frac{(\lambda - q)^2 - q\Sigma + \frac{1}{4}\Sigma^2}{2r^2} - \frac{\lambda r^i \hat{S}^i}{r^3} \qquad i = 1,2,3 \tag{1}$$

where A_i^D is the vector potential for a magnetic monopole of unit strength,

$$\Sigma = \begin{pmatrix} \sigma_3 & 0 \\ 0 & 0 \end{pmatrix} \quad , \quad \hat{S}^i = \begin{pmatrix} 0 & 0 \\ 0 & \frac{\sigma^i}{2} \end{pmatrix} \tag{2}$$

and λ is a free parameter. H_D describes the quantum dynamics of two spin 0 particles and one spin 1/2 particle with electric charge -1/e in the field of dyons with electric charge e and magnetic charges respectively $(q \mp 1/2)/e$ and q/e. We have found that H_D admits an OSp(2,1) spectrum supersymmetry which we used to obtain its spectrum and eigenfunctions[5].

B. In the special case $\lambda = 2q$, the two lower components of H_D read

$$H_1 = H_0 \mathbb{1}_2 - q B^i \sigma_i \qquad B^i = \frac{r^i}{r^3} \tag{3}$$

with

$$H_0 = \frac{1}{2}\left[P_i - q A_i^D\right]^2 - \frac{1}{r} + \frac{q^2}{2r^2} \tag{4}$$

It happens that the spectrum of H_1 possesses high degeneracies. These are understood by viewing H_1 as the supersymmetric partner of $H_0 \mathbb{1}_2$ which is known to have the same spectrum structure as the Coulomb problem (with $q=0$). The constants of motion responsible for the accidental degeneracy of H_1 were obtained and embedded in an $O(4) \oplus U(2/2)$ invariance superalgebra of the combined $H_0 \mathbb{1}_2 \oplus H_1$ system [6]. Their knowledge allowed for an analysis à la Bargmann of the spectrum of H_1 [7].

It is these results that I would like to expand upon in the course of this talk.

A. SPECTRUM SUPERSYMMETRIES OF PARTICLES IN A COULOMB POTENTIAL

In order to derive the spectrum supersymmetries of H_D we shall use dimensional reduction to establish its connection with a 4-dimensional oscillator-like Hamiltonian. The supersymmetries of our 3-dimensional problem will then be inferred from those of this 4-dimensional system. It will be convenient to coordinatize $\mathbb{R}^4$ with 2 complex variables z_a, $a = 1,2$ and their complex conjugate $\overline{z}_a$. We shall denote the corresponding vector fields by $\partial_a = \partial/\partial z_a$, $\overline{\partial}_a = \partial/\partial \overline{z}_a$. Let r_i, $i = 1,2,3$ be the standard Cartesian coordinates on $\mathbb{R}^3$. Dimensional reduction will be effected via the Hopf map :

$$\pi : \mathbb{C}^2 \setminus \{0\} \longrightarrow \mathbb{R}^3 \setminus \{0\}$$
$$r^i = \pi^i(z) = \overline{z}_a \sigma^i_{ab} z_b \qquad i = 1,2,3 \qquad (5)$$

where σ^i stand for the usual Pauli matrices. This projection defines $\mathbb{R}^4 \setminus \{0\}$ as a $U(1)$-bundle over $\mathbb{R}^3 \setminus \{0\}$. (Summation over repeated indices will be understood throughout.)

1. The 4-dimensional system

Consider the supercharges

$$Q = \left(\partial_a - \frac{\lambda \overline{z}_a}{|z|^2} \right) \eta_a \qquad Q^\dagger = \left(-\overline{\partial}_a - \frac{\lambda z_a}{|z|^2} \right) \eta^\dagger_a \qquad (6)$$

with λ a free parameter and the η_s verifying

$$\{\eta_a, \eta_b\} = \{\eta_a^\dagger, \eta_b^\dagger\} = 0 , \quad \{\eta_a, \eta_b^\dagger\} = 2\delta_{ab} \tag{7}$$

We shall use the following realization of this Clifford algebra :

$$\eta_1 = -\frac{1}{\sqrt{2}}\begin{pmatrix} 0 & 1+\sigma_3 \\ 1-\sigma_3 & 0 \end{pmatrix} \quad \eta_2 = -\frac{1}{\sqrt{2}}\begin{pmatrix} 0 & \sigma^1+i\sigma^2 \\ -\sigma^1-i\sigma^2 & 0 \end{pmatrix} \tag{8}$$

The anticommutators involving Q and Q^t are given by

$$\{Q, Q\} = \{Q^\dagger, Q^\dagger\} = 0 \tag{9}$$

and

$$\begin{aligned} H &= \tfrac{1}{2}\{Q, Q^\dagger\} \\ &= -\partial_a \bar{\partial}_a + \frac{\lambda}{|z|^2}(\lambda - C) - 2\lambda \frac{\bar{z}_a \sigma^i_{ab} z_b \hat{S}^i}{|z|^4} \end{aligned} \tag{10}$$

where

$$\begin{aligned} C &= X + \Sigma \\ X &= z_a \partial_a - \bar{z}_a \bar{\partial}_a \quad , \quad \Sigma = \begin{pmatrix} \sigma_3 & 0 \\ 0 & 0 \end{pmatrix} \end{aligned} \tag{11a}$$

and

$$\hat{S}^i = \begin{pmatrix} 0 & 0 \\ 0 & \frac{\sigma^i}{2} \end{pmatrix} \tag{11b}$$

Note that X is the generator of the U(1)-action on the fibers of $\mathbf{R}^4 \setminus \{0\} \to \mathbf{R}^3 \setminus \{0\}$.

Now it is not too difficult to see that we can adjoin to Q, Q^t and H, two more odd generators (S, S^t) and 3 more even generators (D, K, Y) to form an $OSp(2,1)$ realization. Indeed one can check that

$$S = \bar{z}_a \eta_a \qquad S^\dagger = z_a \eta_a^\dagger \qquad \text{(superconformal)} \tag{12a}$$

$$D = \tfrac{1}{2}\left(z_a \partial_a + \bar{z}_a \bar{\partial}_a + 2\right) \qquad \text{(dilations)} \tag{12b}$$

$$K = \bar{z}_a z_a \qquad \text{(conformal)} \tag{12c}$$

$$Y = \tfrac{1}{2}\left(z_a \partial_a - \bar{z}_a \bar{\partial}_a\right) + \Sigma - \lambda \qquad \text{("parity")} \tag{12d}$$

together with Q, Q' and H satisfy the structure relations that characterize the superalgebra OSp(2,1). These are

$$\{Q,Q^\dagger\} = 2H \qquad\qquad \{S,S^\dagger\} = 2K \tag{13a}$$

$$\{Q,S^\dagger\} = -2D - 2\iota Y \qquad \{Q^\dagger, S\} = -2D + 2\iota Y \tag{13b}$$

$$[H,S] = -\iota Q \qquad\qquad [H,S^\dagger] = -\iota Q^\dagger \tag{13c}$$

$$[K,Q] = \iota S \qquad\qquad [K,Q^\dagger] = \iota S^\dagger \tag{13d}$$

$$[D,Q] = -\tfrac{1}{2}Q \quad [D,S] = \tfrac{1}{2}S \quad [D,Q^\dagger] = -\tfrac{1}{2}Q^\dagger \quad [D,S^\dagger] = \tfrac{1}{2}S^\dagger \tag{13e}$$

$$[Y,Q] = \tfrac{1}{2}Q \quad [Y,S] = \tfrac{1}{2}S \quad [Y,Q^\dagger] = -\tfrac{1}{2}Q^\dagger \quad [Y,S^\dagger] = -\tfrac{1}{2}S^\dagger \tag{13f}$$

$$[H,D] = \iota H \quad [H,K] = 2\iota D \quad [D,K] = \iota K \tag{13g}$$

with all the other { } equal to zero.

We remark that all the above charges are invariant under the SU(2)-action generated by

$$J^i = -\tfrac{1}{2}\left(z_a \sigma^i_{ab} \partial_b - \bar{z}_a \sigma^i_{ab} \bar{\partial}_b\right) + \hat{S}^i \qquad i = 1,2,3 \tag{14}$$

We also note that C = 2X + Σ commutes with J^i and with all the OSp(2,1) generators. This observation will play a crucial rôle. In summary, the full symmetry algebra of the 4-dimensional problem that we have just defined is

$$OSp(2,1) \oplus SU(2) \oplus U(1) \tag{15}$$

2. The 3-dimensional system

Let us take the following superalgebra element :

$$R = \tfrac{1}{2}(H + K)$$

$$= \tfrac{1}{2}\left(-\partial_a\bar{\partial}_a + |z|^2 + \frac{\lambda^2 - \lambda C}{|z|^2} - 2\lambda\frac{\bar{z}_a \sigma^i_{ab} z_b \hat{S}^i}{|z|^4}\right) \tag{16}$$

and introduce the eigenvalue equation :

$$R\Psi = (-2E)^{-\frac{1}{2}}\Psi \tag{17}$$

In order to project this equation from $\mathbb{R}^4 \setminus \{0\}$ to $\mathbb{R}^3 \setminus \{0\}$, we shall require that the 4-dimensional wave function Ψ be equivariant under the $U(1)$-action generated by X. More precisely, we shall take Ψ in the $U(1)$ representation with weights

$$\text{diag}\left(q - \frac{\Sigma}{2}\right) = \text{diag}\left(\begin{matrix} q-\frac{1}{2} & & & \\ & q+\frac{1}{2} & & \\ & & q & \\ & & & q \end{matrix}\right) \tag{18}$$

Equivariance under this representation is expressed by the condition $X\Psi = (q - \Sigma/2)\Psi$ or equivalently

$$C\Psi = 2q\Psi \tag{19}$$

Let us point out here that the symmetries of the projected system shall be those of the 4-dimensional system which preserve this constraint. Since C is central, it means that the basis elements of our $OSp(2,1) \oplus SU(2)$ realization all generate symmetries of the 3-dimensional problem.

To carry out the projection it is convenient to introduce the Euler coordinates

$$0 < r \leq \infty \, , \, 0 \leq \theta < \pi \, , \, 0 \leq \phi < 2\pi \, , \, 0 \leq \omega < 4\pi \tag{20}$$

for $\mathbb{R}^4 \setminus \{0\} \sim \mathbb{R}^+ \times S^3 \sim \mathbb{R}^+ \times SU(2)$ in terms of which

$$z_1 = \sqrt{r}\cos\frac{\theta}{2}\exp\left\{\frac{i}{2}(\omega-\phi)\right\} \quad z_2 = \sqrt{r}\sin\frac{\theta}{2}\exp\left\{\frac{i}{2}(\omega+\phi)\right\} \tag{21a}$$

and

$$r_1 = r\cos\phi\sin\theta \quad r_2 = r\sin\phi\sin\theta \quad r_3 = r\cos\theta \tag{21b}$$

In these coordinates the generator X takes the form

$$X = -i\frac{\partial}{\partial\omega} \tag{22}$$

and the solution to equation (19) is given by[*]

$$\Psi(r,\theta,\phi,\omega) = e^{i(q-\frac{1}{2}\Sigma)\omega}\psi(r,\theta,\phi) \tag{23}$$

---

[*] We shall henceforth always designate the 4-dimensional wave function by an upper case psi (Ψ) and the 3-dimensional wave function by a lower case psi (ψ).

Note that q must be an integer or a half-integer for Ψ to be single-valued. Using these wave functions in equation (17), the ω-dependence can be separated out and one is left with the following eigenvalue problem :

$$(\pi_* R)\psi = (-2E)^{-\frac{1}{2}}\psi \qquad (24)$$

where

$$\pi_* R = \frac{1}{2}\left\{ r\left[-i\frac{\partial}{\partial r^i} - (q - \frac{1}{2}\Sigma)A_i^D\right]^2 + r \right.$$

$$\left. + \frac{(\lambda - q)^2 - q\Sigma + \frac{1}{4}\Sigma^2}{r} - \frac{2\lambda r^i S^i}{r^2} \right\} \qquad (25)$$

Multiplying (24) by $1/r$ and rescaling according to $r^i \to (-2E)^{\frac{1}{2}}r^i$ we further find that ψ satisfies the Schrödinger equation

$$H_D\psi = E\psi \qquad (26)$$

with H_D given by equation (1). As already mentioned, this equation describes the quantum dynamics of two spin 0 and one spin 1/2 particles in dyon fields. We have shown that it admits an $OSp(2,1)\oplus SU(2)$ spectrum supersymmetry.

3. The quantum numbers

The eigenstates of H_D can now be obtained by constructing bases for representations of $OSp(2,1)\oplus SU(2)$. One choise of quantum numbers for the basis states of the irreducible representations of $OSp(2,1)\oplus SU(2)$ is provided by the eigenvalues of the Casimir operators associated to the canonical chain of maximal subalgebras:

$$OSp(2,1)\oplus SU(2) \supset O(2)\oplus O(2,1)\oplus U(1) \supset O(2)\oplus O(2)\oplus U(1)$$

$$C_2,C_3 \qquad J^2 \qquad Y \qquad C_0 \qquad J_3 \qquad R$$

The Casimir operator C_0 of $O(2,1)$ is well known and given by

$$C_0 = \frac{1}{2}(HK + KH) - D^2 \qquad (27)$$

The quadratic and cubic Casimir operators of $OSp(2,1)$ have respectively the following expression :

$$C_2 = C_0 - Y^2 + \frac{1}{4}[Q, S^\dagger] + \frac{1}{4}[Q^\dagger, S] \tag{28a}$$

$$C_3 = Y\left(C_2 + \frac{1}{8}[Q, S^\dagger] + \frac{1}{8}[Q^\dagger, S] - \frac{1}{2}\right)$$
$$- \frac{1}{8}[Q, S^\dagger]D + \frac{1}{8}[Q^\dagger, S]D - \frac{1}{8}[Q, Q^\dagger]K - \frac{1}{8}[S, S^\dagger]H \tag{28b}$$

In our realization, C_2 and C_3 are completely determined in terms of j, q and λ so that at fixed angular momentum the system is described by those representations of $OSp(2,1)$ for which

$$C_2 = j(j+1) - j_0(j_0 + 1) \qquad\qquad j_0 = |q| - \frac{1}{2} \tag{29a}$$

$$C_3 = (q - \lambda)[j(j+1) - j_0(j_0 + 1)] \tag{29b}$$

As usual, the eigenvalues of J^2 are written in the form $j(j+1)$, $j - j_0$, $j_0 + 1$, ... and those of J_3 denoted by $m = -j, -j+1, \ldots, j$. Now, in order to characterize the states belonging to these irreducible representations, it is convenient to replace C_0 and Y by

$$\gamma^5 = \begin{pmatrix} 1 & 0 \\ 0 & -1 \end{pmatrix} \tag{30a}$$

and

$$\hat{A} = \frac{1}{2}(1 + \gamma^5)\Sigma + \frac{1}{2}(1 - \gamma^5)\operatorname{sign}(i[Q, S^\dagger] + i[Q^\dagger, S]) \tag{30b}$$

which represent an equivalent pair of labelling operators. It is not difficult to check that C_0 and Y can unambiguously be reconstructed in term of γ^5 and $\hat{A}$ which both have their eigenvalues (χ and $\hat{\alpha}$) equal to ± 1. As a matter of fact

$$C_0 = \left\{ \left[J^2 - j_0(j_0 + 1) + (q - \lambda)^2 \right]^{\frac{1}{2}} - \frac{1}{4}(1 - \gamma^5)\hat{A} \right\}^2 - \frac{1}{4} \tag{31a}$$

$$Y = \frac{1}{4}(1 + \gamma^5)\hat{A} + (q - \lambda) \tag{31b}$$

Finally, from the representation theory of $O(2,1)$, we know the eigenvalues of R to be given by

$$r_n = (\Delta_{j,\hat{\alpha},\chi} + n) \qquad\qquad n = 0, 1, 2 \tag{32}$$

if those of C_0 are written in the form

$$C_0 = \Delta_{j,\hat{\alpha},\chi}(\Delta_{j,\hat{\alpha},\chi} - 1) \tag{33}$$

Here,

$$\Delta_{j,\hat{\alpha},\chi} = \left[j(j+1) - j_0(j_0+1) + (q-\lambda)^2 \right]^{\frac{1}{2}} - \frac{1}{4}(1-\chi)\hat{\alpha} + \frac{1}{2} \tag{34}$$

In summary, we have the following eigenvalue equations to characterize the states of our system :

$$J^2 \,|\, j,m,\hat{\alpha},\chi,n \rangle = j(j+1)\,|\, j,m,\hat{\alpha},\chi,n \rangle \tag{35a}$$

$$J_3 \,|\, j,m,\hat{\alpha},\chi,n \rangle = m\,|\, j,m,\hat{\alpha},\chi,n \rangle \tag{35b}$$

$$\hat{A} \,|\, j,m,\hat{\alpha},\chi,n \rangle = \hat{\alpha}\,|\, j,m,\hat{\alpha},\chi,n \rangle \tag{35c}$$

$$\gamma^5 \,|\, j,m,\hat{\alpha},\chi,n \rangle = \chi\,|\, j,m,\hat{\alpha},\chi,n \rangle \tag{35d}$$

$$R \,|\, j,m,\hat{\alpha},\chi,n \rangle = (\Delta_{j,\hat{\alpha},\chi}+n)|\, j,m,\hat{\alpha},\chi,n \rangle \tag{35e}$$

4. The OSp(2,1) representations

The action of the remaining $OSp(2,1)$ generators on the $|\, j, m, \hat{\alpha}, \chi, n \rangle$ state vectors has been obtained recently [6]. This is most easily achieved by going to a Cartan-type basis for $OSp(2,1)$. Introduce the following ladder operators

$$B_\pm = \frac{1}{2}\left[K - H \pm 2iD \right] \tag{36a}$$

$$F_\pm^L = -\frac{1}{2}\left[iS \pm Q \right] \quad , \quad F_\mp^R = (F_\pm^L)^\dagger \tag{36b}$$

The $OSp(2,1)$ structure relations then become

$$[R, B_\pm] = \pm B_\pm \qquad [B_+, B_-] = -2R \tag{37a}$$

$$\{F^{L,R}, F^{L,R}\} = 0 \qquad \{F_\pm^L, F_\pm^R\} = B_\pm \qquad \{F_\pm^L, F_\mp^R\} = R \pm Y \tag{37b}$$

$$[R, F_\pm^{L,R}] = \pm\frac{1}{2}F_\pm^{L,R} \quad [Y, F_\pm^L] = -\frac{1}{2}F_\pm^L \quad [Y, F_\pm^R] = \frac{1}{2}F_\pm^R \tag{37c}$$

$$[B_\pm, F_\pm^{L,R}] = 0 \qquad [B_\pm, F_\mp^{L,R}] = \mp F_\pm^{L,R} \tag{37d}$$

After a little work, one find that $B_\pm$ and $F_\pm^{(L,R)}$ act as follows on our basis states:

$$B_\pm|j,m;\hat\alpha,\chi,n\rangle = [(\Delta_{j,\hat\alpha,\chi}+n)(\Delta_{j,\hat\alpha,\chi}+n\pm1)$$
$$-\Delta_{j,\hat\alpha,\chi}(\Delta_{j,\hat\alpha,\chi}-1)]^{\frac12}|j,m;\hat\alpha,\chi,n\pm1\rangle \tag{38a}$$

$$F_\pm^L|j,m;\hat\alpha,-1,n\rangle =$$
$$\hat\alpha\,\mathcal{Q}_{\hat\alpha}[(1\pm\hat\alpha)\Delta_{j,\hat\alpha,\chi}+\tfrac12(1-\hat\alpha)+n]^{\frac12}|j,m;-1,1,n\pm\tfrac12-\tfrac{\hat\alpha}{2}\rangle \tag{38b}$$

$$F_\pm^R|j,m;\hat\alpha,-1,n\rangle =$$
$$\mathcal{Q}_{-\hat\alpha}[(1\pm\hat\alpha)\Delta_{j,\hat\alpha,\chi}+\tfrac12(1-\hat\alpha)+n]^{\frac12}|j,m;1,1,n\pm\tfrac12-\tfrac{\hat\alpha}{2}\rangle \tag{38c}$$

$$F_\pm^L|j,m;1,1,n\rangle =$$
$$\mathcal{Q}_-[(1\mp1)\Delta_{j,\hat\alpha,\chi}+\tfrac12\pm\tfrac12+n]^{\frac12}|j,m;1,-1,n\pm\tfrac12+\tfrac12\rangle \tag{38d}$$
$$+\mathcal{Q}_+[(1\pm1)\Delta_{j,\hat\alpha,\chi}+\tfrac12\pm\tfrac12+n]^{\frac12}|j,m;-1,-1,n\pm\tfrac12-\tfrac12\rangle$$

$$F_\pm^R|j,m;1,1,n\rangle = 0 \tag{38e}$$
$$F_\pm^L|j,m;-1,1,n\rangle = 0 \tag{38f}$$

$$F_\pm^R|j,m;-1,1,n\rangle =$$
$$\mathcal{Q}_+[(1\mp1)\Delta_{j,\hat\alpha,\chi}+\tfrac12\pm\tfrac12+n]^{\frac12}|j,m;1,-1,n\pm\tfrac12+\tfrac12\rangle \tag{38g}$$
$$-\mathcal{Q}_-[(1\pm1)\Delta_{j,\hat\alpha,\chi}+\tfrac12\pm\tfrac12+n]^{\frac12}|j,m;-1,-1,n\pm\tfrac12-\tfrac12\rangle$$

where

$$\mathcal{Q}_{\hat\alpha} = \left\{\frac{[\Delta_{j,\hat\alpha,\chi}+\hat\alpha(q-\lambda)]}{2\Delta_{j,\hat\alpha,\chi}}\right\}^{\frac12} \tag{39}$$

By going to a coordinate realization, solving for the ground state and applying repeatedly yhe ladder operators, the wave functions can then be obtained simply (see reference [6]).

5. The spectrum of H_D

The spectrum of H_D can now be straight forwardly gotten. We have arranged our equations for E to be the eigenvalue of H_D. Now from eqs. (17) and (35e) we have

$$R = (-2E)^{-\frac{1}{2}} = (\Delta_{j,\hat{\alpha},\chi} + n) \tag{40}$$

It trivially follows that

$$E_{n,j,\hat{\alpha},\chi} = \frac{-1}{2(\Delta_{j,\hat{\alpha},\chi} + n)^2} \tag{41}$$

In the special case $\lambda = 2q$ the two lower components $(\chi = -1)$ of H_D become

$$H' = \frac{1}{2}\vec{v}^2 - \frac{1}{r} + \frac{q^2}{2r^2} - eB^i\sigma^i \tag{42}$$

with

$$\begin{aligned}
\vec{v} &= (\vec{p} - e\vec{A}) & \vec{A} &= g\vec{A}^D \\
\vec{B} &= \vec{\nabla} \times \vec{A} = g\frac{\vec{r}}{r^3} & q &= eg
\end{aligned} \tag{43}$$

We readily see that H' is the Pauli Hamiltonian that describes the dynamics of a spin-1/2 particle with gyromagnetic ratio equal to 4 and electric charge equal to e in the presence of a $1/r^2$-potential and in the field of a dyon with electric charge $-1/e$ and magnetic charge g. It is immediate to check that when $\lambda = 2q$:

$$\Delta_{j,\hat{\alpha},\chi=1} = j - \frac{1}{2}\hat{\alpha} + 1 \tag{44}$$

and thus,

$$E'_{n,j,\hat{\alpha}} = \frac{-1}{2\left(n + j - \frac{1}{2}\hat{\alpha} + 1\right)^2} = \frac{-1}{2p^2} \tag{45}$$

$$p = q, q+1, \ldots$$

Interestingly enough, H' manifestly exhibits accidental degeneracies similar to those of the Coulomb problem. Indeed when $p \neq q$ we find a $2(p^2-q^2)$-fold degeneracy at the level p. (In the ground state $p = |q|$ and the $2|q|$-fold degeneracy is accounted for by the rotational invariance of the problem.) We shall discuss the hidden symmetries that explain these accidental degeneracies in the second part of this talk.

B. HIDDEN SYMMETRIES OF A SPINNING PARTICLE IN A DYON FIELD

In order to study the invariance of

$$H' = H_0 - q\frac{r^i}{r^3}\sigma^i \tag{46a}$$

where

$$H_0 = \frac{1}{2}\vec{v}^2 - \frac{1}{r} + \frac{q^2}{2r^2} \tag{46b}$$

it is convenient to form the following 4 × 4 block diagonal Hamiltonian which simultaneously describe two spin 0 and one spin 1/2 particles:

$$H = \begin{bmatrix} (H_0 + \frac{1}{2q^2})\mathbb{1}_2 & 0 \\ 0 & (H' + \frac{1}{2q^2}) \end{bmatrix} \tag{47a}$$

$$= \frac{1}{2}[\vec{v}^2 + U^2(r)] - P_L q\frac{r^i}{r^3}\Sigma^i \tag{47b}$$

In equations (47),

$$U(r) = \frac{q}{r} - \frac{1}{q} \tag{48a}$$

$$\Sigma^i = \begin{pmatrix} \sigma^i & 0 \\ 0 & \sigma^i \end{pmatrix} \quad , \quad P_{\substack{R \\ L}} = \frac{1}{2}(1 \pm \gamma_5) \tag{48b}$$

1. Symmetries of H (generalization of the Runge-Lenz vector)

Since the symmetries of H include the symmetries of H' as a subset, we may as well look for these operators that commute with H. The rotational invariance of H leads evidently to the conservation of the angular momentum

$$J^i = L^i + \frac{1}{2}P_L\Sigma^i \qquad [J^i, J^j] = i\epsilon^{ijk}J^k \tag{49a}$$

$$L_i = \epsilon_{ijk}r^j v^k - q\frac{r^i}{r} \tag{49b}$$

Now the spin 0 Hamiltonian H_0 is known to admit the following constant of motion :

$$\Lambda_0^i = \tfrac{1}{2}\varepsilon^{ijk}\left(v^j L^k - L^j v^k\right) - \frac{r^i}{r}$$

$$= r^i v^2 - v^i r^j v^j + \frac{q}{r}L_i + \frac{q^2 r^i}{r^2} - \frac{r^i}{r} \tag{50}$$

We extended [7] this result by showing that H commutes with

$$\Lambda^i = \Lambda_0^i \mathbb{1}_4 - P_L\{\varepsilon^{ijk}\Sigma^j v^k - \frac{q}{r}\Sigma^i + \frac{q^2 r^i r^j}{r^3}\Sigma^j + \frac{1}{2q^2}\Sigma^i\} \tag{51}$$

thus providing a generalization of the Runge-Lenz vector to a system with spin. Λ_i clearly transforms as a vector :

$$[J^i, \Lambda^j] = \iota\varepsilon^{ijk}\Lambda^k \tag{52}$$

and quite surprisingly, upon commuting Λ_i with itself, we see that there is yet another vector quantity Ω_i which is conserved. Indeed,

$$[\Lambda^i, \Lambda^j] = -\iota\varepsilon^{ijk}[2(H - \frac{1}{2q^2})J^k + \Omega^k] \tag{53}$$

with

$$\Omega^i = P_L[\tfrac{1}{2}\Sigma^i v^2 - \Sigma^i v^j v^i + \varepsilon^{ijk}U(r)\Sigma^j v^k - \tfrac{1}{2}U^2(r)\Sigma^i] \tag{54}$$

Note that every term in Ω_i is affected by a spin matrix. Note also that $P^R\Sigma^i$ trivially commutes with H since H_0 is spin independent. Now a problem remains unsolved : Why have we added a term $1/2q^2$ to the Hamiltonian ?I The answer to this question will become apparent in the next section. Let us just say as a hint that with this addition the spectrum of H' precisely starts at zero and is positive.

2. Supersymmetries of H

At this point, we would like to obtain the commutation relations satisfied by the symmetries of H' : $P^L J_i$, $P^L\Lambda_i$, $P^L\Omega_i$. The evaluation of these commutators can be greatly simplified by making use of the fact that H actually possesses an invariance superalgebra. First, one can check that the supercharges

$$Q_1 = \frac{1}{\sqrt{2}}(\iota\gamma^i v^i + \gamma^0 U(r)) \tag{55a}$$

$$Q_2 = \iota\gamma^5 Q_1 \tag{55b}$$

satisfy

$$\{Q_\alpha, Q_\beta\} = 2\delta_{\alpha\beta}H \qquad \alpha, \beta = 1, 2 \tag{56}$$

(We use a standard chiral representation for the Dirac matrices.) Since Q_α and Λ_i both commute with H, so must their commutator and indeed we find that $[\Lambda_i, Q_\alpha]$ gives two new conserved vector supercharges. These are

$$q_\alpha^i = 2iq\,[\Lambda^i, Q_\alpha] \tag{57}$$

with

$$q_1^i = \frac{1}{\sqrt{2}}(\gamma^0 v^i + i\epsilon^{ijk}\gamma^j v^k - i\gamma^i U(r)) \tag{58a}$$

$$q_2^i = i\gamma^5 q_1^i \tag{58b}$$

Note also that

$$[P_R \Sigma^i, Q_\alpha] = -i q_\alpha^i \tag{59}$$

We can now form the quaternion

$$\mathbb{Q}_\alpha = Q_\alpha \mathbb{1}_2 + i q_\alpha^i \sigma^i \tag{60}$$

to see that the supercharges transform according to a spin $1/2$ representation :

$$[J^i, \mathbb{Q}_\alpha] = \tfrac{1}{2}\mathbb{Q}_\alpha \sigma^i \tag{61}$$

This is in contradistinction with the situation encountered in part **A** where the supercharges were rotation scalars.

3. Structure relations

We are now in a position to describe the structure relations satisfied by all the above invariant operators. In order to arrive at these relations, it suffices to evaluate the anticommutators of odd generators and the commutators between odd and even operators. Once these quantities are known, the Lie product on the "bosonic" subalgebra can easily be deduced. Since the conserved supercharges are of first order in the velocities, the procedure for determining the commutator of $P^L J_i$, $P^L\Lambda_i$ and $P^L\Omega_i$ is clearly simpler than a brute force calculation.

For the sake of illustration, we shall take $H \neq 0$. (Although the ground state needs to be considered separately, its treatment is completely analogous to what will be presented below.) Let us define

$$M_1^i = \frac{1}{2}\left(J^i + \frac{1}{2H}\Omega^i + \beta\left(-q\Lambda^i + \frac{1}{2H}\Omega^i\right)\right) \tag{62a}$$

$$M_2^i = \frac{1}{2}\left(J^i + \frac{1}{2H}\Omega^i - \beta\left(-q\Lambda^i + \frac{1}{2H}\Omega^i\right)\right) \tag{62b}$$

$$M_3^i = -\frac{1}{2H}\Omega^i \tag{62c}$$

$$M_4^i = \frac{1}{2}P_R\Sigma^i \tag{62d}$$

with

$$\beta = \left(1 - 2q^2H\right)^{-\frac{1}{2}} \tag{63}$$

In terms of these operators, one obtains the following set of structure relations for our constants of motion :

$$\{Q_\alpha, Q_\beta\} = 2\delta_{\alpha\beta}H \quad \{Q_\alpha, Q_\beta^i\} = -4H\epsilon_{\alpha\beta}(M_3^i + M_4^i) \tag{64a}$$

$$\{Q_\alpha^i, Q_\beta^j\} = 2\delta_{\alpha\beta}\delta^{ij}H - 4\epsilon_{\alpha\beta}\epsilon^{ijk}H(M_3^k - M_4^k) \tag{64b}$$

$$[J^i, Q_\alpha] = \frac{1}{2}Q_\alpha\sigma^i \quad [M_1^i, Q_\alpha] = [M_2^i, Q_\alpha] = 0 \tag{64c}$$

$$[M_3^i, Q_\alpha] = \frac{1}{2}Q_\alpha\sigma^i \quad [M_4^i, Q_\alpha] = -\frac{1}{2}\sigma^iQ_\alpha \tag{64d}$$

$$[J^i, J^j] = \iota\epsilon^{ijk}J^k \quad [J^i, M_a^j] = \iota\epsilon^{ijk}M_a^k \tag{64e}$$

$$[M_a^i, M_b^j] = \iota\delta_{ab}\epsilon^{ijk}M_a^k \tag{64f}$$

$$[\gamma^5, Q_\alpha] = -2\iota Q_\alpha \quad [\gamma^5, J^j] = [\gamma^5, M_a^i] = 0 \tag{64g}$$

$$a = 1,2,3,4$$

Let us observe that :

i) At fixed energy, the operators J^i, M_a^i, Q_α and γ^5 form a closed set under the graded product. The superalgebra thus obtained is identified as being SU(2/2)⊕SU(2)⊕SU(2).

ii) The symmetries of the Hamiltonian H', namely $\{$ $^{PL}J_i$, $^{PL}\Lambda_i$, $^{PL}\Omega_i$ $\}$ or equivalently $\{$ $^{PL}M_1{}^i$, $^{PL}M_2{}^i$, $^{PL}M_3{}^i$ $\}$ form by themselves the algebra $SU(2) \oplus SU(2) \oplus SU(2)$.

4. Spectrum analysis à la Bargmann

We will now apply the Pauli-Bargmann method to show that the spectrum of H' is given by

$$E = \frac{-1}{2p^2} \qquad\qquad p = q, q+1, \ldots \tag{65}$$

In the process, the $2(p^2-q^2)$-fold degeneracy of the p-level will be completely accounted for. Here also we shall restreint to $H \neq 0$.

From the definition of the operators $M_a{}^i$, we have

$$M_1^2 = \frac{1}{4}\left\{ J^2 + \frac{1}{H}(1+\beta) + \frac{1}{4H^2}(1+\beta)^2\Omega^2 \right.$$
$$\left. + \beta^2 q^2 \Lambda^2 - \frac{q}{H}\beta(1+\beta)\Lambda\cdot\Omega - 2q\beta J\cdot\Lambda \right\} \tag{66a}$$

$$M_2^2 = M_1^2(\beta \rightarrow -\beta) \tag{66b}$$

$$M_3^2 = \frac{1}{4H^2}\Omega^2 \tag{66c}$$

With a little bit of effort, it is possible to show that

$$\Omega^2 = 3H^2 \tag{67a}$$

$$J\cdot\Omega = AH + \frac{2}{q}J\cdot\Lambda - \frac{1}{2}H + \frac{1}{2q^2} - 2 \tag{67b}$$

$$\Lambda\cdot\Omega = -2(J\cdot\Lambda)H + \frac{1}{q}J\cdot\Omega + 2qH + \frac{3}{2q}H \tag{67c}$$

$$\Lambda^2 = 2H(J^2 - q^2 + \frac{1}{4}) - \frac{1}{q^2}J^2 + 2HA + 2HA$$
$$+ 2H + \frac{2}{q}J\cdot\Lambda \tag{67d}$$

with

$$A = \sum^i L^i + 2q \sum^i \frac{r^i}{r} + 1 \tag{68}$$

Substituting in equation (66), one then finds

$$M_a^2 = \frac{1}{4}(s_a - 1)(s_a + 1) \qquad a = 1,2,3 \tag{69}$$

with

$$S_1 = q\left[\left(1-2q^2H\right)^{-\frac{1}{2}} - 1\right] \tag{70a}$$

$$S_2 = q\left[\left(1-2q^2H\right)^{-\frac{1}{2}} + 1\right] = S_1 + 2q \tag{70b}$$

$$S_3 = 2 \tag{70c}$$

From the representation theory of SU(2), we know that these S_a are the dimensions of the three SU(2) representations generated by $M_a{}^i$, $a = 1, 2, 3$. From eq. (70a), it is immediate to obtain

$$H = \frac{1}{2q^2} - \frac{1}{2(S_1 + q)^2} \tag{71}$$

Setting

$$S_1 = (p - q) \tag{72}$$

and substracting the $1/2q^2$ that had been added, we therefore find

$$H' = \frac{-1}{2p^2} \qquad p = q+1, q+2, \ldots \tag{73}$$

As for the degeneracies, these are given by the product of the dimensions of the corresponding three SU(2) representations. For the level p, we have

$$S_1 \cdot S_2 \cdot S_3 = S_1 \cdot (S_1 + 2q) \cdot 2 = (p-q)(p+q) \cdot 2 = 2(p^2 - q^2) \tag{74}$$

We should point out that the degeneracies of the ground state can similarly be analyzed (see ref. [8]).

Acknowledgements

One of us (L.V.) would like to thank H.J. de Vega, N. Sanchez and R. Kerner for their kind invitation to give a talk in their seminar series. We also want to ackowledge the help of M. Mayrand in the preparation of the manuscript.

This work has been supported in part through funds provided by U.S. Department of Energy (D.O.E.), the Natural Science and Engineering Research Council (NSERC) of Canada and the Quebec Ministry of Education.

References

[1]   For a review see E. D'Hoker, V.A. Kostelecký and L. Vinet in "Dynamical Groups and Spectrum Generating Algebras" edited by A. Barut, A. Bohm and Y. Ne'eman, World Scientific (to appear).

[2]   E. D'Hoker and L. Vinet, Phys. Lett, 137B, 72 (1984).

[3]   E. D'Hoker and L. Vinet, Lett. Math. Phys., 8, 439 (1984).

[4]   E. D'Hoker and L. Vinet, Commun. Math. Phys., 97, 391 (1985).

[5]   E. D'Hoker and L. Vinet, Nucl. Phys., B260, 79 (1985).

[6]   L. Benoit, M.Sc. Thesis (Université de Montréal) unpublished (1986). L. Benoit and L. Vinet, A supermultiplet in a Coulomb potential : states and wave functions from the representation theory of OSp(2,1), LPNUM preprint (to appear).

[7]   E. D'Hoker and L. Vinet, Phys.Rev.Lett., 55, 1043 (1985).

[8]   E. D'Hoker and L. Vinet, Lett. Math. Phys., 12, 71 (1986).

CLASSICAL r-MATRICES, LAX EQUATIONS,
POISSON LIE GROUPS AND DRESSING TRANSFORMATIONS

M.A. Semenov-Tian-Shansky[*) **)]

A B S T R A C T

We discuss the theory of Poisson Lie groups
which provides a natural framework for the study
of integrable Hamiltonian systems on a lattice
and of the dressing transformations in soliton
theory.

Classical r-matrices are the semiclassical counterparts of quantum R-matrices originally introduced by R. Baxter and fully exploited within the frames of the Quantum Inverse Scattering Method developed by L.D. Faddeev and his disciples. The notion of a classical r-matrix was originally proposed by E.K. Sklyanin [1]. A wide range of classical r-matrices was then classified by A.A. Belavin and V.G. Drinfel'd [2] as a step in a classification of quantum R-matrices. To describe the relevance of classical r-matrices for the study of classical integrable systems recall that they are usually characterized by the following properties which seem at the first glance unrelated.

(a) There is a natural Poisson bracket on the phase space and the equations of motion are Hamiltonian.

(b) All integrals of motion arise as spectral invariants of a linear operator. They are in involution with respect to the Poisson bracket.

(c) The solution of the equations of motion may be reduced to some kind of Riemann-Hilbert problem.

As was first noticed by the present author [3] the r-matrix approach provides a natural link between these properties. The simplest theorem of this kind deals with the case when the Poisson brackets are linear, i.e. the phase space is realized as a Poisson submanifold of a dual space to a certain Lie algebra. This case is most commonly known in the literature. For this case the theorem may be regarded as a slight

---
[*)] Ecole Normale Supérieure - Paris
[**)] On leave of absence from the Steklov Mathematical Institute, Leningrad - USSR
Talk given at the Paris-Meudon Seminar Series.

generalization of the Kostant-Adler-Symes commutativity theorem /4/.
However, the r-matrix approach naturally leads to a very interesting
new class of Poisson brackets that are non-linear. This class of
Poisson brackets (eventually referred to as the Sklyanin brackets, cf.
/1/) is particularly well suited to the study of integrable systems
on a lattice arising naturally as lattice approximations of continuous
integrable models, or the semiclassical versions of quantum magnetics.
V.G.Drinfel'd /5/ was the first to notice that Sklyanin's original
definition leads to an important new geometric concept, that of a
Poisson Lie group.[+)].

The geometry of Poisson Lie groups has an interest of its own. In particular we are naturally led to generalize the notions of Hamiltonian
reduction, the moment map etc. So far, several applications of these
concepts were already indicated :

(a) The description of the symplectic leaves of Poisson Lie groups
due to Drinfel'd.

(b) The geometrical theory of Lax systems on the lattice, due to the
present author.

(c) The theory of dressing transformations.

We shall give a few comments on the latter point. Recall that dressing
transformations were introduced by V.E.Zakharov and A.B.Shabat /6/
(However, they have made no emphasize on their group theoretic properties). In a different setting they were rediscovered in a now famous
series of papers of the Kyoto scientists /7/. An excellent geometrical
treatment of the subject is given in /8/,/9/. The puzzling property
of dressing transformations is that they do not respect the Poisson
brackets on the phase space. It is particularly interesting to understand this phenomenon in order to decide whether the dressing transformation group survives quantization. Normally this would have been
the case if they preserved the Poisson brackets. Since they fail to do
so, the situation becomes obscure. As it happens, the theory of Poisson

----

[+)] I prefer to rectify the original term introduced by Drinfel'd which
was "Hamilton Lie groups". This latter term seems ambiguous since
the natural actions of such groups are not Hamiltonian.

Lie groups provides a clue for a natural characterization of the behaviour of Poisson brackets under the dressing transformations /10/. Almost all the technique available by now is used for the proof, so it
seems to be a good example to introduce the reader to the domain which
I believe is still promising.

The present lectures are organized as follows.

We start with the definition of classical r-matrices and the proof of
the generalized Kostant-Adler-Symes theorem. Applications of this
theorem to particular dynamical systems are now quite numerous (see
/11/, /12/, /13/), so we shall not dwell upon this. Instead, we display
two examples of r-matrices that are less widely known, the rational
and the elliptic ones. Our exposition combines ideas borrowed from
/3/ and /4/. It is based on my recent joint paper with A.G.Reyman
/15/. These examples actually will not be used in the sequel, so the
reader may skip them. An extension to systems depending on a continuous
x parameter is briefly indicated.

Our second major theme is the definition of Poisson Lie groups. The
main results here are due to V.G.Drinfel'd /5/. We proceed then to the
study of Lax equations on the lattice and the Poisson properties of
dressing transformations. The latter subject was already exposed in
/10/. In order to diminish the overlap I shall consider here mainly
the continuous case rather than the discrete one.

Acknowledgements . Conversations with V.G.Drinfel'd during the past
years were extremely important for me. I also wish to express my
sincere gratitude to Prof.L.Breen and Prof.J.L.Verdier and to the
Ecole Normale Supérieure for their hospitality.

§1. DOUBLE LIE ALGEBRAS AND THE GENERALIZED KOSTANT-ADLER-SYMES THEOREM

1.1. **Generalities**. Throughout the paper we shall deal with the
Poisson brackets. Recall that by definition (cf. /16/) the Poisson
bracket on a smooth manifold $\mathcal{M}$ is the structure of a Lie algebra on
$C^\infty(\mathcal{M})$ which satisfies the Leibnitz rule

$$(1.1) \quad \{\varphi_1, \varphi_2 \varphi_3\} = \{\varphi_1, \varphi_2\}\varphi_3 + \{\varphi_1, \varphi_3\}\varphi_2.$$

The degenerate brackets are not excluded. By definition, functions
lying in the center of $C^\infty(\mathcal{M})$ are called Casimir functions. A
mapping $f : \mathcal{M} \to \mathcal{N}$ of two Poisson manifolds is called a Poisson

mapping if

(1.2) $\{\varphi \cdot f, \psi \cdot f\}_{\mathcal{M}} = \{\varphi, \psi\}_{\mathcal{N}} \cdot f$ for any $\varphi, \psi \in C^{\infty}(\mathcal{N})$

A submanifold V of $\mathcal{M}$ is called a Poisson submanifold if there is a
Poisson structure on V such that the natural embedding $V \hookrightarrow \mathcal{M}$ is a
Poisson mapping. Such a structure is clearly unique if it exists. Every
point $x \in \mathcal{M}$ is contained in a minimal Poisson submanifold which carries
a symplectic structure. These symplectic manifolds form a stratifica-
tion of $\mathcal{M}$.

The best known example of a Poisson bracket is the Lie Poisson bracket
defined on the dual space to a Lie algebra $\mathcal{Y}$. By definition,

(1.3) $\{\varphi, \psi\}(L) = L\left([d\varphi(L), d\psi(L)]\right), \quad \varphi, \psi \in C^{\infty}(\mathcal{G}^{*})$.

(Obviously, $d\varphi(L) \in (\mathcal{G}^{*})^{*} = \mathcal{Y}$). The symplectic leaves of the
Lie Poisson bracket coincide with the orbits of the adjoint Lie group
in the space $\mathcal{Y}^{*}$. By an abuse of language we shall simply call them
$\mathcal{Y}$ - orbits.

Other examples of Poisson structures will be given in §3 .

1.2. Definition of double Lie algebras.

Definition 1.1 : Let $\mathcal{Y}$ be a Lie algebra, $R \in \text{End}\,\mathcal{Y}$ a linear operator acting on $\mathcal{Y}$. We shall say that R defines the structure of a
double Lie algebra on $\mathcal{Y}$ if the bracket

(1.4) $[X, Y]_{R} = \frac{1}{2}[RX, Y] + \frac{1}{2}[X, RY]$

satisfies the Jacobi identity. In that case there are two structures
of a Lie algebra on the same linear space $\mathcal{Y}$. The Lie algebra $\mathcal{Y}$
equipped with the Lie bracket $[\,;\,]_{R}$ will be denoted $\mathcal{Y}_{R}$. Consequently,
there are two Poisson brackets on the dual space $\mathcal{Y}^{*}$, the Lie Poisson
brackets of the algebras $\mathcal{Y}$, $\mathcal{Y}_{R}$
We denote by $I(\mathcal{Y})$ the space of Casimir functions of $\mathcal{Y}$.

As is well known, $I(\mathcal{Y})$ coincides precisely with the space of $ad^{*}\mathcal{Y}$-
-invariants in $C^{\infty}(\mathcal{Y}^{*})$. The following theorem serves as a main
motivation of the definition above.

Theorem 1.1. (i) Elements of $I(\mathcal{Y})$ are in involution with respect to
the Lie-Poisson bracket of $\mathcal{Y}_{R}$.
      (ii) Let $h \in I(\mathcal{Y})$. The equation of motion defined by h
with respect to the Lie-Poisson bracket of $\mathcal{Y}_{R}$ has the generalized

Lax form

$$(1.5) \qquad \frac{dL}{dt} = - ad^*_{\mathcal{Y}} M.L \quad , \quad M = \frac{1}{2} R (dh(L)) \quad , \quad L \in \mathcal{Y}^*$$

Proof. (i) Let $h_1, h_2 \in I(\mathcal{Y})$ and put $X_i = dh_i(L)$, $L \in \mathcal{Y}^*$. By definition ,

$$\{h_1, h_2\}(L) = L([X_1, X_2]_R) = \frac{1}{2} L([RX_1, X_2] + [X_1, RX_2])$$
$$= \frac{1}{2} ad_{\mathcal{Y}} X_2 . L R(X_1) - \frac{1}{2} ad^*_{\mathcal{Y}} X_1 . L(R X_2)$$

Now, since the functions h_i are $ad^* \mathcal{Y}$ - invariant both terms vanish.

(ii) By definition, the hamiltonian equation of motion defined by a function $h \in C^\infty(\mathcal{Y})$ may be written in the form

$$(1.6) \qquad \frac{dL}{dt} = - ad^*_{\mathcal{Y}_R} dh(L).L$$

From (1.4) it is clear that

$$(1.7) \qquad ad^*_{\mathcal{Y}_R} X.L = \frac{1}{2} ad^*_{\mathcal{Y}} RX.L + R^*(ad^*_{\mathcal{Y}} X.L)$$

Since $h \in I(\mathcal{Y})$ the second term vanishes and we arrive to (1.5).
The geometrical meaning of theorem 1.1 is fairly simple. Recall that in the present case there are two systems of orbits in $\mathcal{Y}^*$, those of $\mathcal{Y}$ and of $\mathcal{Y}_R$. Theorem 1.1 says essentially that hamiltonian equations of motion with Hamiltonians $h \in I(\mathcal{Y})$ respect both. Indeed, formulae (1.5), (1.6) mean that the velocity vector is always tangent to the intersection of a $\mathcal{Y}$-orbit and a $\mathcal{Y}_R$-orbit of $L \in \mathcal{Y}^*$. In many cases of interest these intersections coincide precisely with the Liouville tori for our Hamiltonian systems which implies complete integrability (cf. /11/) .
So far we have not specified the conditions on R imposed by demanding that (1.4) satisfy the Jacobi identity. We choose to impose the following sufficient condition which is called the (modified) classical Yang-Baxter identity:

$$(1.8) \qquad [RX, RY] = R([RX, Y] + [X, RY]) - [X, Y]$$

for any $X, Y \mathcal{Y}$. It is straight forward to check that (1.8) implies the Jacobi identity for (1.4). Indeed for any $X, Y, Z \in \mathcal{Y}$

$$4[[X, Y]_R, Z]_R + c.p. = [X, [RY, RZ] - R([RY, Z] + [Y, RZ])] + c.p.$$
$$= - [X, [Y, Z]] - c.p. = 0.$$

The motivation for the choice of condition (1.8) which is of course stronger than merely demanding the Jacobi identity to be valid is provided by the global version of Theorem 1.1. which we are going to formulate. We start with the following simple statement.

Proposition 1.1. (i) Suppose $R \in \text{End } \mathcal{Y}$ satisfies (1.8) then the mapping

(1.9) $\iota_R : \mathcal{Y} \rightarrow \mathcal{Y} \oplus \mathcal{Y} : X \longmapsto \left(\frac{1}{2} (RX + X), \frac{1}{2} (RX - X) \right)$

is a Lie algebra monomorphism.

(ii) Any $X \in \mathcal{Y}$ admits a unique decomposition

(1.10) $$X = X_+ - X_-$$

with $(X_+ , X_-) \in \text{Im } \iota_R$.

Proof. We write $X_{\pm} = \frac{1}{2} (RX \pm X)$. Clearly , $X_+ - X_- = X$ which implies (ii) . In particular, i_R is a monomorphism. The formula $\left([X, Y]_R \right)_{\pm} = [X_{\pm}, Y_{\pm}]$ is a direct consequence of (1.8) .

Let $G(G_R)$ be a connected simply connected Lie group with the Lie algebra $\mathcal{Y}(\mathcal{Y}_R)$. Extend i_R to a Lie group monomorphism $i_R : G_R \rightarrow G \times G$ Let m : $G_R \rightarrow G$ be the composition map $x \longmapsto (x_+ , x_-) \longmapsto x_+ x_-^{-1}$

Proposition 1.2. (i) The image of m is an open cell in G.

(ii) For $x \in \mathcal{J}_m$ the decomposition

(1.11) $x = x_+ x_-^{-1}$, $(x_+ , x_-) \in \mathcal{J}_m \, i_R$

is unique.

We are now able to state the global version of Theorem 1.1.

Theorem 1.2.   Let $h \in I(\mathcal{Y})$. For $L \in \mathcal{Y}^*$ put $X_L = dh(L)$. Let $x_+(t) \, x_-(t)$ be the solutions to the factorization problem (1.11.) with the left-hand side given by

(1.12) $x(t) = \exp t \, x_L$

The integral curve of the equation (1.5) with the Hamiltonian h starting at $L \in \mathcal{Y}^*$ is given by

(1.13) $L(t) = A_d^* \, x_{\pm}(t)^{-1} \cdot L.$

We shall outline the proof that will be eventually generalized to the case of Poisson Lie groups considered in §5.

Let $T^* G$ be the cotangent bundle of G . There are two natural actions of G on $T^* G$ by left and right translations. Both actions are Hamiltonian with respect to the standard symplectic structure. Hence there are two moment maps $\mu_l, \mu_r : T^* G \rightarrow \mathcal{Y}^*$. The following statement is of course well known (cf. /16/).

Proposition 1.2.   Equip $\mathcal{Y}^*$ with the Lie Poisson bracket and let $'\mathcal{Y}^*$ be another copy of $\mathcal{Y}^*$ equipped with the opposite Poisson bracket

Then

$$\mu_\ell \swarrow \quad T^*G \quad \searrow \mu_r$$
$$\mathfrak{a\jmath}^* \qquad\qquad {}'\mathfrak{a\jmath}^*$$

(1.14)

is a dual pair of Poisson mappings.

In particular, we may regard μ_r as a Hamiltonian reduction map with respect to the group action which corresponds to μ_ℓ and vice versa. Now choose $h \in I(\mathfrak{g})$ and extend it to a function $\hat{h} \in C^\infty(T^*_G)$ which is both right- and left-G-invariant.

Lemma 1. Fix a trivialization of $T^*G \simeq G \times \mathfrak{g}^*$ by means of left translations. The Hamiltonian flow on T^*G defined by $\hat{h}$ is given by the formula

(1.15) $\quad (x, L) \longmapsto (x\, e^{t\,dh(L)}\ L)\quad , \quad (x, L) \in G \times \mathfrak{a\jmath}^*$.

Proof. In a trivialization chosen the moment maps μ_ℓ, μ_r are given by

(1.16) $\quad \mu_\ell : (x, L) \longmapsto Ad^*_x . L \quad , \quad \mu_r : (x, L) \longmapsto - L$.

Now, consider the action of the group G_R on G given by

(1.17) $\quad h : x \to h_+ \, x \, h_-^{-1} \quad , \quad h \in G_R , \quad x \in G$

where $h \mapsto h_\pm$ are the standard homomorphisms $G_R \to G$ defined above. This action clearly extends to a Hamiltonian action on $T^*G \simeq G \times \mathfrak{g}^*$.

Lemma 2. (i) The mapping $T^*G \to \mathfrak{g}^*$ given by

(1.18) $\quad (x, L) \longmapsto Ad^*_{x_-} . L \quad , \qquad x = x_+ \, x_-^{-1}$,

defines a cross-section of the action (1.17) over the open cell $G_+ . G_- \times \mathfrak{g}^* \subset T^*G$.

(ii) The quotient Poisson structure on $\mathfrak{a\jmath}^*$ coincides with the Lie Poisson bracket of $\mathfrak{g}_R$.

(iii) The reduced Hamiltonian on $\mathfrak{a\jmath}^*$ corresponding to $\hat{h}$ is h itself. The quotient Hamiltonian flow is given by (1.13). The proof of (i), (iii) is obvious. To prove (ii) let us notice that since $m : G_R \to G$ is an immersion it naturally extends to a symplectic map $m^* : T^*G_R \to T^*G$. This mapping intertwins the action (1.17) with the standard action $G_R \times T^*G_R$ by left translations. Hence (iii) is corollary of Proposition 1.2 .

Note. The reader will eventually see that most of these arguments extend to the case of Poisson Lie groups. This does not apply to the last argument, so we shall replace it by a straightforward computation (which is fairly possible in the just considered case as well).

§ 2. EXAMPLES

2.1. The following class of r-matrices satisfying (1.8) is by far the most important. Suppose $\mathcal{y}_+, \mathcal{y}_- \subset \mathcal{y}$ are Lie subalgebras such that

(2.1) $$\mathcal{y} = \mathcal{y}_+ + \mathcal{y}_-$$

as a linear space. Let $P_\pm$ be the projection operator onto $\mathcal{y}_\pm$ parallel to the complementary subalgebra . Put

(2.2) $$R = P_+ - P_- .$$

The check of (1.8) is straightforward. In this particular case Theorem 1.1. coincides with the Kostant-Adler-Symes commutativity theorem.

We now give some examples of decompositions (2.1) which lead to numerous applications.
Let $\mathcal{y}$ be a simple Lie algebra, and let $\mathcal{L}(\mathcal{y}) = \mathcal{y} \otimes \mathbb{C}[\lambda^{-1}, \lambda]$ be its loop algebra. Fix a graduation

(2.3) $$\mathcal{L}(\mathcal{y}) = \bigoplus_{j \in \mathbb{Z}} \mathcal{L}_j$$

and put $$\mathcal{L}(\mathcal{y})_+ = \bigoplus_{j \geq 0} \mathcal{L}_j , \quad \mathcal{L}(\mathcal{y})_- = \bigoplus_{j < 0} \mathcal{L}_j .$$
Clearly,

(2.4) $$\mathcal{L}(\mathcal{y}) = \mathcal{L}(\mathcal{y})_+ + \mathcal{L}(\mathcal{y})_- ,$$

so any graduation gives rise to an r-matrix of the form (2.2).
Let us indicate in passing an example of an r-matrix satisfying (1.8) though not of the form (2.2). Suppose that (2.3) is a principal graduation and put $k_\pm = \bigoplus_{j > 0} \mathcal{L}_{\pm j}$, $h = \mathcal{L}_o$. Clearly, f is a Cartan subalgebra in $\mathcal{y}$. Let $P_\pm$, P_o be the projection operators onto $k_\pm$, f in the decomposition
$$\mathcal{L}(\mathcal{y}) = k_+ + h + k_-$$
Proposition 2.1. Fix $\Theta \in \mathrm{End}\, h$ such that $\det(1 - \Theta) \neq 0$. Then

(2.5) $$R\Theta = P_+ - P_- + (1 + \Theta)(1 - \Theta)^{-1} P_o$$

satisfies (1.8) .
Note that (2.5) differs from an operator of the form (2.2) by a finite dimensional perturbation. A theorem due to Belavin and Drinfel'd asserts that all graded r-matrices on loop algebras are essentially of this type. See /2/, /3/ for more information on the subject.

2.2. There is an extension of the decomposition (2.5) suggested by Cherednik /14/. For $\nu \in \mathbb{CP}_1 = \mathbb{C} \cup \{\infty\}$ let λ_ν be the local parameter i.e. $\lambda_v = \lambda - v$, $v \neq \infty$, $\lambda_\infty = 1/\lambda$. Put $a_{y_v} = a_y \otimes \mathbb{C}[\lambda_v^{-1}, \lambda_v]]$

For a fixed finite set $\mathscr{D} \subset \mathbb{CP}_1$ put

(2.6)
$$a_{y_{\mathscr{D}}} = \bigoplus_{v \in \mathscr{D}} a_{y_v} \ .$$

Put
(2.7)
$$a_{y_v}^+ = a_y \otimes \mathbb{C}[[\lambda_v]] \ , \ v \neq \infty, \ a_{y_\infty}^+ = a_y \otimes_{\lambda_\infty} \mathbb{C}[[\lambda_\infty]]$$

and let $a_y(\mathscr{D})$ be the Lie algebra of rational functions with values in a_y which are regular outside $\mathscr{D}$. Let $i : a_y(\mathscr{D}) \to a_{y_{\mathscr{D}}}$ the natural embedding which assigns to each function the set of its Laurent series at points $v \in \mathscr{D}$.

Proposition 2.2.   There is a direct sum decomposition

(2.8)
$$a_{y_{\mathscr{D}}} = a_y(\mathscr{D}) + a_{y_{\mathscr{D}}}^+$$

This is merely a reformulation of the well known Mittag-Leffler theorem for $\mathbb{CP}_1$.

Hence we may associate with (2.8) an **r**-matrix of the form (2.2). This leads to the so-called Lax equations with rational spectral parameter.

2.3. As the construction above suggests, we may try to replace $\mathbb{CP}_1$ by an algebraic curve of higher genus. However, in this case there is an obstruction to a decomposition of the form (2.8), as it follows from the Mittag-Leffler theorem for curves. We shall briefly indicate how this obstruction may be avoided for elliptic curves. This leads to the so-called elliptic r-matrices /1/, /2/ .

Recall first the so-called Heisenberg representation.

Let
$$I_1 = \text{diag} \ (1, \varepsilon, \ldots, \varepsilon^{n-1}) \ , \qquad \varepsilon = e^{2\pi i/n} \ ,$$

(2.9) $I_2 = \begin{pmatrix} 0 & 1 & & 0 \\ & 0 & 1 & \ddots \\ & & & \ddots & 1 \\ 1 & & \ldots & & 0 \end{pmatrix}$

For a $= (a_1 \cdot a_2) \in \mathbb{Z}_n \times \mathbb{Z}_n$ put $I_a = I_1^{a_1} I_2^{a_2}$. The assignement $a \mapsto I_a$ defines a projective representation of $\mathbb{Z}_n \times \mathbb{Z}_n$:

(2.10) $I_a I_\ell = I_{\ell + a} I_a \ e^{\frac{\pi i}{n} \langle a, b \rangle}$, $\langle a, b \rangle = a_2 b_1 - a_1 b_2$

The associated representation of $\mathbb{Z}_n \times \mathbb{Z}_n$ in End $\mathbb{C}^n$ is equivalent to the regular representation. One has, obviously,

(2.11) $I_a I_b I_a^{-1} = e^{2\pi i/n \langle a, b \rangle} I_b$.

Now, let $a_y = \text{sl}(n, \mathbb{C})$. Let Γ be an elliptic curve,
$\Gamma = \mathbb{C}/\mathbb{Z}\omega_1 + \mathbb{Z}\omega_2$. We identify $\mathbb{Z}_n \times \mathbb{Z}_n$ with the subgroup $\Gamma_n \subset \Gamma$ of points of order n. Fix a finite set $\mathscr{D} \subset \Gamma$ which does not contain

any points which are equivalent modulo Γ_n . Let $\mathcal{E}(\mathcal{D})$ be the Lie algebra of rational functions on Γ which are regular outside $\Gamma_n \cdot \mathcal{D}$ and satisfy the automorphy condition

(2.12)
$$X(\lambda + a) = I_a \, X(\lambda) \, I_a^{-1}.$$

As above, let

$$\mathfrak{y}_\mathcal{D} = \bigoplus_{\nu \in \mathcal{D}} \mathfrak{y}_\nu \qquad , \qquad \mathfrak{y}_\nu = \mathfrak{y} \otimes \mathbb{C}[\lambda_\nu^{-1}, \lambda_\nu],$$
$$\mathfrak{y}_\mathcal{D}^+ = \bigoplus_{\nu \in \mathcal{D}} \mathfrak{y}_\nu^+ \qquad , \qquad \mathfrak{y}_\nu^+ = \mathfrak{y} \otimes \mathbb{C}[[\lambda_\nu]].$$

Clearly, $\mathcal{E}(\mathcal{D})$ is naturally embebbed into $\mathfrak{y}_\mathcal{D}$.

Proposition 2.3.   There is a direct sum decomposition

(2.12)
$$\mathfrak{y}_\mathcal{D} = \mathcal{E}(\mathcal{D}) + \mathfrak{y}_\mathcal{D}^+.$$

Proof . By the Mittag-Leffler theorem a rational function on Γ with the prescribed principal parts X_ν at the points $\nu \in \Gamma_n \cdot \mathcal{D}$ exists if and only if

(2.13)
$$\sum_{\nu \in \Gamma_n \cdot \mathcal{D}} \text{Res}_\nu \, X_\nu \, d\lambda = 0$$

This is immediate from (2.11) since tr $X_\nu = 0$. Hence the decomposition (2.12) exists. The uniqueness follows from the irreducibility of Heisenberg's representation.

Let φ_a be the unique rational function which has only simple poles at Γ_n , satisfies the functional equation

(2.14)
$$\varphi_a(\lambda + b) = e^{2\pi i/n \langle a, b \rangle} \, \varphi_a(\lambda)$$

and is normalized by the condition $\text{Res.} \, \varphi_a \, d\lambda = 1$.
Put

(2.15)
$$r(\lambda, \mu) = \sum_{\substack{a \in \Gamma_n \\ a \neq o}} \varphi_a(\lambda - u) \, I_{-a} \otimes I_a$$

Proposition 2.4.   The projection operator onto $\mathcal{E}(\mathcal{D})$ parallel to $\mathfrak{y}_\mathcal{D}^+$ is given by the formula

(2.16)
$$(P X)(\lambda) = \sum_{\nu \in \mathcal{D}} \text{Res}_\nu \, r(\lambda, \mu) \, X_\nu \, d\mu .$$

where the linear operator $I_{-a} \otimes I_a$ is acting by $X \longrightarrow I_{-a} \, \text{tr}(X I_a)$.

Proposition 2.4. establishes a link between the formalism which is used in /1/, /2/ and our present approach. Originally the r-matrices were considered as functions with the values in $\mathfrak{y} \otimes \mathfrak{y}$. In our approach we associate with such functions linear operators. The ellip-

tic r-matrix (2.15) was indeed (for n = 2) the first example of a
classical r-matrix ever studied /1/.

2.4   Let us now indicate how the present formalism may be used to
produce integrable Lax equations. We proceed in several steps.

(a) The pairing

$$(2.17) \quad \langle X, Y \rangle = \sum_{v \in \mathcal{D}} \text{Res}_v \; \text{tr} \; X_v \; Y_v \; d\lambda$$

is non-degenerate and allows to identify $\mathcal{Y}_\mathcal{D}$ with its dual. Note also
that $(\mathcal{Y}_\mathcal{D}^+)^* \simeq \mathcal{E}(\mathcal{D})$. Another model for the dual to $\mathcal{Y}_\mathcal{D}^+$ is provided
by $\mathcal{Y}_\mathcal{D}^- = \bigoplus_{v \in \mathcal{D}} \mathcal{Y} \otimes \lambda_v^{-1} \mathbb{C}[\lambda_v^{-1}]$. The two models are related by a map
which assigns to a rational function on Γ satisfying (2.12) the set
of its principal parts at $v \in \mathcal{D}$. The Poisson submanifolds in the
space $(\mathcal{Y}_\mathcal{D}^+)^*$ are easy to describe. In particular, we have

Proposition 2.4. Functions with simple poles at $v \in \mathcal{D}$ form a Poisson
submanifold of $\mathcal{E}(\mathcal{D}) \simeq (\mathcal{Y}_\mathcal{D}^+)^*$. The symplectic leaves lying in it
coincide with the coadjoint orbits of $\prod^{\text{card}\mathcal{D}} G$, $G = SL(n, \mathbb{C})$

(b)   Let $I(\mathcal{Y})$ be the algebra of Ad G-invariants on $\mathcal{Y} = sl(n, \mathbb{C})$
The algebra of Casimir functions of $\mathcal{Y}_\mathcal{D}$ is generated by the functionals of the form

$$(2.18) \quad h_{\alpha, \varphi} : L \mapsto \sum_{r \in \mathcal{D}} \text{Res}_v \; \varphi_v (L_v) \; \alpha_v \; d\lambda \; ,$$

where $\varphi_v \in I(\mathcal{Y})$, $\alpha_v \in \mathbb{C}[\lambda_v^{-1}, \lambda_v]]$, $L = (L_v)_{v \in \mathcal{D}} \in \mathcal{Y}_\mathcal{D}$
By restricting these functionals to the orbits described above we get
Hamiltonians in involution giving rise to Lax equations of the form

$$(2.19) \quad \frac{dL}{dt} = [L, M] \; , \quad L, M \in \mathcal{E}(\mathcal{D}) \; , \; M = P^o (d \, h_{\alpha, v}(L))$$

They are usually referred to as Lax equations with the spectral parameter on an elliptic curve. By applying Theorems 1.1, 1.2, we may
systematically construct such equations and their solutions (cf./15/).

2.5. The examples considered so far give rise to finite dimensional
systems admitting Lax representations $dL/dt = [L, M]$, where
L, M are matrices possibly depending on spectral parameter. In many
cases it is natural to assume that L, M also depend on a spatial
variable x. Lax equation then takes the form $\frac{dL}{dt} = \frac{\partial M}{\partial x} + [L, M]$.
There is a natural way to include such equations into the present
formalism.We explain it in brief since it will be of importance in the

study of dressing transformations (see § 6. below).

Let $\mathcal{G}$ be a Lie algebra with an invariant scalar product. It will be convenient to assume that $\mathcal{G}$ is a matrix algebra. We denote by G the corresponding matrix Lie group. For the time being the reader may assume that $\mathcal{G}$ is finite dimensional. However, in realistic applications $\mathcal{G}$ is always a loop algebra (see below). Put $\mathcal{G} = C^\infty(\mathbb{R}/\mathbb{Z}, \mathcal{G})$, $\mathbf{G} = C^\infty(\mathbb{R}/\mathbb{Z}, G)$. Suppose R ∈ End $\mathcal{G}$ satisfies the Yang-Baxter identity (1.8). We extend it to $\mathcal{G}$ by setting $(RX)(x) = R(X(x))$. Let $\mathcal{G}_R$ be the corresponding Lie algebra with the Lie bracket (1.4).

There is a 2-cocycle on $\mathcal{G}$ defined by

(2.20) $\quad \omega(X, Y) = \int dx\, (X, \partial_x Y).$

Let $\hat{\mathcal{G}}$ be the central extension of $\mathcal{G}$ defined by the cocycle (2.20). Put

(2.21) $\quad \omega_R(X, Y) = \frac{1}{2}\omega(X, Y) + \frac{1}{2}\omega(X, RY).$

Proposition 2.5. (i) Formula (2.21) defines a 2-cocycle on $\mathcal{G}_R$
(ii) Let $\hat{\mathcal{G}}_R$ be the corresponding central extension of $\mathcal{G}_R$
Then $(\hat{\mathcal{G}}, \hat{\mathcal{G}}_R)$ is a double Lie algebra.

It is particularly nice when the operator R is skew with respect to the inner product on $\mathcal{G}$. In that case $\omega_R = 0$, so the central extension $\hat{\mathcal{G}}_R$ splits. Hence the orbits of $\mathcal{G}_R$ and $\hat{\mathcal{G}}_R$ coincide, they are clearly "continuous products" of orbits of $\mathcal{G}$.

Since in the sequel we shall be dealing almost entirely with this case, it is worth giving a formal definition.

Definition 2.1 A double Lie algebra $(\mathcal{G}, \mathcal{G}_R)$ is called a Baxter Lie algebra if (i) the operator R ∈ End $\mathcal{G}$ satisfies the modified Yang-Baxter identity (1.8) ; (ii) there is a (fixed) invariant inner product on $\mathcal{G}$ and R is skew with respect to it.

Let us now describe the Casimir functions on $\hat{\mathcal{G}}$.
Proposition 2.6. Let us identify $\hat{\mathcal{G}}$ with its dual by means of the inner product

(2.22) $\quad (X, Y) = \int dx\, (X(x), Y(x)),$

so that $\hat{\mathcal{G}}^* \simeq \mathcal{G} \oplus \mathbb{C}$. The coadjoint action of $\hat{\mathcal{G}}$ on $\hat{\mathcal{G}}^*$ is given by

(2.22) $\quad ad^*_{\hat{\mathcal{G}}} X \cdot (L, e) = ([X, L] + e\, \partial_x X, 0),$
$$L \in \mathcal{G}, e \in \mathbb{C}.$$

It integrates to the action of G given by

(2.23) $\quad Ad^* \, g \cdot (L, e) = \left(g \, L \, g^{-1} + e \, \partial_x \, g g^{-1}, e \right).$

Notice that (2.23) coincides with the gauge transformations which are connected with the linear differential equation

(2.24) $\qquad e \, \partial_x \, \psi - L \psi \qquad , \quad \psi \in C^\infty \, (\mathbb{R}, G).$

Let ψ_L be the fundamental solution to (2.24) normalized by the condition

(2.25) $\quad \psi_L (0) - I \cdot$ $\qquad$ (the identity matrix)

By definition, the monodromy matrix $T(L) = \psi_L(1)$.

Theorem 2.1. (Floquet). Two points (L,e), (L',e') $(e \neq 0)$ lie on the same coadjoint orbit in $\hat{\mathcal{G}}^*$ if and only if $e' = e$ and the matrices $T(L)$, $T(L')$ are conjugate in G'.

Corollary . The Casimir functions on $\hat{\mathcal{G}}^*$ are of the form $L \to \psi(T(L))$ where $\psi \in C^\infty(G)$ is a central function.

Note. It is clear now that the codimension of orbits in $\hat{\mathcal{G}}^*$ is equal to rank $\mathcal{G} = \ell$. There are also precisely ℓ generators of the algebra of Casimir functions on each hyperplane $e = \text{const} \neq 0$ in $\mathcal{G}^*$. Hence to get sufficiently many integrals of motion provided by theorem 1.1. we must assume $\ell = \infty$. This is the case when $\mathcal{G}$ is a loop algebra.

Theorem 2.1. shows in particular that our geometric approach incorporates the conventional inverse spectral transform methods which are based on the study of the auxiliary linear problem (2.24) .

An extremely important point is the study of Poisson properties of the monodromy map which we now state.

For $\psi \in C^\infty(G)$ let $\nabla_\psi, \nabla_\psi'$ be its left and right gradients. By definition $\nabla_\psi, \nabla_\psi' \in \mathcal{G}$ and

(2.26) $\left(\nabla_\psi(x), X \right) - \left(\frac{d}{dt} \right)_{t=0} \psi \left(e^{tX} x \right) , \quad \left(\nabla_\psi'(x), X \right) = \left(\frac{d}{dt} \right)_{t=0} \psi \left(x e^{tX} \right).$

Theorem 2.2. Let $\psi_1, \psi_2 \in C^\infty(G)$. The Poisson bracket of the functionals $L \to \psi_i(T(L))$ is given by

(2.27) $\left\{ \psi_1 \circ T, \, \psi_2 \circ T \right\}_R (L) = \frac{1}{2} \left(R X_1', \, X_2' \right) - \frac{1}{2} \left(R(X_1), \, X_2 \right),$
$\qquad X_i = \nabla_{\psi_i}(T(L)) , \qquad X_i' = \nabla_{\psi_i}'(T(L)).$

In particular, the right hand side satisfies the Jacobi identity.

We shall give a proof of a more general formula in §6 . For the time being (2.27) will serve us as a motivation for the following definition.

Definition 2.3. Suppose G is a Lie group and that there is a structure of a Baxter Lie algebra on its Lie algebra $\mathcal{Y}$. The Poisson bracket on G is defined by

$$(2.28) \quad 2\{\varphi_1, \varphi_2\} = (R(X_1'), X_2') - (R(X_1), X_2), \quad X_i = \nabla\varphi_i, \quad X_i' = \nabla'\varphi_i .$$

Formula (2.28) will be referred to as the Sklyanin bracket on G .

Corollary . Equip G with the bracket (2.28). The monodromy

$$T: \mathcal{Y}_R^* \to G: L \to T(L) \qquad \text{is a Poisson mapping.}$$

We shall study the properties of (2.28) directly in §3 .

Proposition 2.7. Multiplication in G induces a Poisson mapping

$$G \times G \to G .$$

Although the direct proof is quite simple we shall give a 'physical proof' based on theorem 2.2. Observe that functions $L \in \mathcal{Y}$ supported on a small patch $I_1 \subset \mathbb{R}/\mathbb{Z}$ form a Poisson subspace in $\mathcal{Y}_R^*$, $\mathcal{Y}^{I_1}$ say. If $I^1 \cap I^2 = \varnothing$, then $\mathcal{Y}^{I_1 \cup I_2} = \mathcal{Y}^{I_1} \times \mathcal{Y}^{I_2}$ as Poisson manifolds. Finally, if $L = L_1 + L_2$, $L_i \in \mathcal{Y}^{I_i}$, then $T(L) = T(L_1) T(L_2)$. Now our claim follows from (2.27) .

Proposition 2.7 was indeed the key motivation to introduce the notion of Poisson Lie groups. We shall return to the study of equation (2.24) in §6 and proceed now to formal definitions.

§3 . POISSON LIE GROUPS. DEFINITIONS AND PROPERTIES.

3.1. The following definition was already anticipated in the proof of proposition 2.7.

Definition 3.1. The product of two Poisson manifolds $\mathcal{M}_1, \mathcal{M}_2$ is the manifold $\mathcal{M}_1 \times \mathcal{M}_2$ equipped with the Poisson bracket

$$(3.1) \quad \{\varphi, \psi\}_{\mathcal{M}_1 \times \mathcal{M}_2} (x, y) = \{\varphi(\cdot, y), \psi(\cdot, y)\}_{\mathcal{M}_1}(x) + \{\varphi(x, \cdot), \psi(x, \cdot)\}_{\mathcal{M}_2}(y).$$

In other words, (3.1) is the unique Poisson structure on $\mathcal{M}_1 \times \mathcal{M}_2$ such that (i) natural projections $P_1 : M_1 \times M_2 \to M_1$, $P_2 : M_1 \times M_2 \to M_2$ are Poisson mappings (ii) $\{ P_1^* \varphi_1 , P_2^* \varphi_2 \} = 0$ for any $\varphi_1 \in C^\infty (M_1)$, $\varphi_2 \in C^\infty (M_2)$.

Definition 3.2. Poisson Lie group is a Lie group equipped with a Poisson bracket such that (i) multiplication in G defines a Poisson mapping $G \times G \to G$ (ii) inversion $x \mapsto x^{-1}$ changes the sign of the Poisson bracket.

Examples 3.1. (i) Any Lie group equipped with a zero Poisson bracket satisfies the axioms. (ii) Let $\mathcal{Y}^*$ be the dual space to a Lie algebra equipped with the Lie Poisson bracket. We regard it as abelian group. Then the axioms are also satisfied.

A much less trivial example is provided by (2.28). We shall come up to its study later.

Let λ_x, $\mathcal{S}_x$ be the left and right translation operator on $C^\infty (G)$ by an element $x \in G$: $\lambda_x \varphi(y) = \varphi(xy)$, $\mathcal{S}_x \varphi(y) = \varphi(xy)$. Multiplication in G induces a Poisson mapping $G \times G \to G$ if

$$(3.2) \quad \{ \varphi, \psi \} (xy) = \{ \lambda_x \varphi, \lambda_x \psi \}(y) + \{ \mathcal{S}_x \varphi, \mathcal{S}_y \psi \}(x) , \quad \varphi, \psi \in C^\infty (G)$$

Recall that any Poisson bracket is bilinear in the derivatives of functions. It is convenient to write down a Poisson bracket on a Lie group in the right- or left-invariant frame. Define the left and right differentials of a function $\varphi \in C^\infty (G)$ by the formula

$$(3.3) \quad \langle \mathcal{D} \varphi(x), X \rangle = \left(\frac{d}{dt} \right)_{t=0} \varphi (e^{tX} x) , \quad \langle \mathcal{D}'\varphi, X \rangle = \left(\frac{d}{dt} \right)_{t=0} \varphi (e^{tX}),$$

$X \in \mathcal{Y}$, $\mathcal{D}\varphi(x)$, $\mathcal{D}'\varphi(x) \in \mathcal{Y}^*$.

Let us define the Hamiltonian operators $\eta, \eta' : G \to \text{Hom} (\mathcal{Y}^*, \mathcal{Y})$ which correspond to our bracket by setting

$$(3.4) \quad \{ \varphi, \psi \}(x) = \langle \eta(x) \mathcal{D}\varphi, \mathcal{D}\psi \rangle = \langle \eta'(x) \mathcal{D}'\varphi, \mathcal{D}'\psi \rangle.$$

Proposition 3.1. Suppose G is a Poisson Lie group. Then functions η, η' satisfy the functional equations

$$(3.5) \quad \begin{aligned} \eta(xy) &= \text{Ad } x . \eta(y) . \text{Ad}^* x^{-1} + \eta(x) , \\ \eta'(xy) &= \text{Ad } y^{-1} . \eta'(x) . \text{Ad}^* y + \eta'(y) \end{aligned}$$

Proof . Obviously,

$$\mathcal{D}\lambda_x \varphi(y) = \text{Ad}^* x^{-1} \mathcal{D}\varphi (xy) , \quad \mathcal{D}'\lambda_x \varphi(y) = \mathcal{D}'\varphi(xy),$$

$$(3.6) \quad \mathcal{D} \, \mathcal{S}_y \, \varphi(x) = \mathcal{D} \, \varphi(xy) \, , \qquad \mathcal{D}' \mathcal{S}_y \, \varphi(x) = Ad^*_y \cdot \mathcal{D}' \varphi(xy)$$

Clearly, (3.2), (3.6), (3.4) imply (3.5)

3.2. Functional equations (3.5) imply in particular that $\eta(e) = \eta'(e) = 0$, hence the Poisson structure on G is always degenerate at the unit element. By linearizing the Poisson bracket at the point e we get the <u>tangent Lie bialgebra</u> of G . To be more precise, fix $\xi_1, \xi_2 \in \mathcal{Y}$ and choose $\varphi_1, \varphi_2 \in C^\infty(G)$ such that $\mathcal{D} \varphi_i(e) = \xi_i$. Put

$$(3.7) \quad [\, \xi_1, \xi_2 \,]_* = \mathcal{D} \{ \varphi_1, \varphi_2 \} (e).$$

Proposition 3.2. Formula (3.7) defines the structure of a Lie algebra on $\mathcal{Y}^*$.

Proof . Formulae (3.4), (3.7) imply that

$$(3.8) \quad [\, \xi_1 \, \xi_2 \,]_* = \langle d\eta(e)\xi_1, \xi_2 \rangle,$$

hence the definition is unambiguous. The Jacobi identity for (3.7) is obvious.

Definition 3.3. Let $\mathcal{Y}$ be a Lie algebra, $\mathcal{G}^*$ its dual. Suppose there is a Lie algebra structure on $\mathcal{Y}^*$, i.e. a mapping $\mathcal{G}^* \wedge \mathcal{G}^* \longrightarrow \mathcal{G}^*$ satisfying the Jacobi identity. The Lie brackets on $\mathcal{Y}$ and $\mathcal{Y}^*$ are said to be consistent if the dual map $\varphi: \mathcal{Y} \to \mathcal{Y} \otimes \mathcal{Y} \simeq Hom(\mathcal{Y}^*, \mathcal{Y})$ is a 1-cocycle on $\mathcal{Y}$, i.e.

$$(3.9) \quad \varphi([X, Y]) = ad \, X \cdot \varphi(Y) - \varphi(Y) \circ ad^* X - ad \, Y \circ \varphi(X) + \\ + \varphi(X) \cdot ad^* Y$$

A pair $(\mathcal{Y}, \mathcal{Y}^*)$ with consistent Lie brackets is called a Lie bialgebra. Lie bialgebras form a category in which the morphisms are such mappings p : $\mathcal{Y} \to h$ that both p and $p^* : h^* \longrightarrow \mathcal{Y}^*$ are Lie algebra homomorphisms.

Theorem 3.1. /5/ (i) Formula (3.7) defines the structure of a Lie bialgebra on $(\mathcal{Y}, \mathcal{Y}^*)$. (We shall refer to it as the tangent Lie bi-algebra of G). (ii) Conversely, let $(\mathcal{Y}, \mathcal{Y}^*)$ be a Lie bialgebra, G a Lie group corresponding to $\mathcal{Y}$. There is a unique Poisson Lie group structure on $\mathcal{Y}$ such that its tangent Lie bialgebra is $(\mathcal{Y}, \mathcal{Y}^*)$. (iii) The correspondance between the Poisson Lie groups and Lie bialgebras is functorial.

Sketch of a proof. Let η be the Hamiltonian operator which corresponds to the Poisson bracket on $\mathcal{G}$. From (3.8) we get

$$\varphi(x) = \left(\frac{d}{dt}\right)_{t=0} \eta\left(e^{tX}\right) .$$

Now, (3.5) implies (3.9). To prove the converse statement we must integrate the 1-cocycle on aj to a 1-cocycle on G. Observe that if φ is trivial i.e. $\varphi(x) = ad\, X \cdot r - r \circ ad^* X$ where $r \in \mathrm{Hom}\,(\mathcal{G}^*, \mathcal{G})$ is a fixed element, the corresponding 1-cocycle on G is given by the obvious formula

$$\eta(x) = Ad\, x \circ r \circ Ad^* x^{-1} - r$$

Operator r is also called classical r-matrix. We shall see below that the general case may be reduced to this special one (cf. the note following Theorem 3.3. below)

3.3. As the reader may have already noticed, there is a difference between the notions of double Lie algebras and Lie bialgebras. Since the former was motivated by applications to integrable systems and the latter by the geometry of Poisson Lie groups, it seems natural to combine the two. An appropriate class lying in the intersection are precisely the Baxter Lie algebras already introduced in §2 (Definition 2.2.).

Proposition 3.3. Suppose $(\mathcal{G}, \mathcal{G}_R)$ is a Baxter Lie algebra. The isomorphism $\mathcal{G}^*$ $\mathcal{G}$ induced by the invariant scalar product on $\mathcal{G}$ equips $(\mathcal{G}, \mathcal{G}^*)$ with the structure of a Lie bialgebra.

Proof. The cocycle φ is in this case trivial and equals

$$\varphi(x) = ad\, X \circ R - R \circ ad\, X .$$

We now come up to the study of Sklyanin brackets (already defined in §2) and their generalizations.

Let G be a Lie group with a Lie algebra $\mathcal{G}$. Assume there is a fixed invariant inner product $\langle \bullet, \bullet \rangle$ on $\mathcal{G}$ and identify $\mathcal{G}^*$ with $\mathcal{G}$ by its means. Accordingly we shall use the notation $\nabla_\varphi, \nabla'_\varphi$ as in (2.26) for left and right gradients of functions on G. Let $R, R' \in \mathrm{End}\ \mathcal{G}$ be skew and satisfy (1.8) .

Proposition 3.4. Formula

(3.10) $\{\varphi, \psi\}_{R,R'} = \frac{1}{2} \langle R(\nabla_\varphi), \nabla_\psi \rangle + \frac{1}{2} \langle R'(\nabla'_\varphi), \nabla'_\psi \rangle$

defines a Poisson structure on G.

Sketch of a proof. The obstruction for the Jacobi identity to be valid
is a tri-linear form in the gradients. Consider first the left- and
right-invariant brackets

$$\{\varphi, \psi\}^{\wedge} = \frac{1}{2} \langle R(\nabla_\varphi), \nabla_\psi \rangle, \quad \{\varphi, \psi\}^{\ell} = \frac{1}{2} \langle R(\nabla'_\varphi), \nabla'_\psi \rangle$$

The corresponding obstructions are given by

$$\Omega^{\wedge}(\nabla_{\varphi_1}, \nabla_{\varphi_2}, \nabla_{\varphi_3}) = \langle \nabla_{\varphi_1}, [R(\nabla_{\varphi_2}), R(\nabla_{\varphi_3})]\rangle + \text{cyclic permutation.}$$

(3.11)

$$\Omega^{\ell}(\nabla'_{\varphi_1}, \nabla'_{\varphi_2}, \nabla'_{\varphi_3}) = - \langle \nabla'_{\varphi_1}, [R\nabla'_{\varphi_2}), R(\nabla'_{\varphi_3})]\rangle + \text{cyclic permutation.}$$

The sign difference is caused by the fact that the Lie algebras of
left- and right-invariant vector fields on G are anti-isomorphic. By
virtue of the Yang-Baxter identity (1.8) the right hand side in (3.11)
simplifies to give

$$\Omega^{\wedge}(\nabla_{\varphi_1}, \nabla_{\varphi_2}, \nabla_{\varphi_3}) = - \langle \nabla_1, [\nabla_{\varphi_2}, \nabla_{\varphi_3}]\rangle,$$
$$\Omega^{\ell}(\nabla'_{\varphi_1}, \nabla'_{\varphi_2}, \nabla'_{\varphi_3}) = + \langle \nabla'_{\varphi_1}, [\nabla'_{\varphi_2}, \nabla'_{\varphi_3}]\rangle.$$

Since the inner product on $\mathcal{G}$ is invariant, these obstructions cancel.

Specifically, the brackets

$$(3.12) \pm \{\varphi, \psi\}_{R, \pm R} = \frac{1}{2} \langle R(\nabla_\varphi), \nabla_\psi \rangle \pm \frac{1}{2} \langle R(\nabla'_\varphi), \nabla'_\psi \rangle$$

satisfy the Jacobi identity. The bracket $\{ , \}_{R,-R}$ is the Sklyanin
bracket already introduced in (2.28).

 The Hamiltonian operators which correspond to (3.10) in the right-
invariant (left-invariant) frame are given by

$$(3.13)\; \eta_{R, R'}(x) = \frac{1}{2} R + \frac{1}{2} \text{Ad } x . R' . \text{Ad } x^{-1}, \quad \eta'_{R,R'}( ) = \frac{1}{2} \text{Ad } x^{-1}. R . \text{Ad } x + \frac{1}{2} R'.$$

In particular, $\eta_{R,-R}$ is obviously a trivial 1-cocycle on G with the
values in End $\mathcal{G}$. Hence we immediately get

Proposition 3.5 . The Sklyanin bracket defines the structure of a
Poisson Lie group on G .

An indirect proof of this statement was already given in §2 . We shall
denote the group G equipped with the bracket (3.10) by $G(R,R')$.

Let us now study the behaviour of the Poisson bracket (3.10) under the
left and right transformations. First, we introduce the following
important definition:

Definition 3.4 . Let G be a Poisson Lie group, $\mathcal{M}$ a Poisson manifold

An action $G \times \mathcal{M} \to \mathcal{M}$ is called a Poisson group action if it is a Poisson mapping, the space $G \times \mathcal{M}$ being equipped with the product Poisson structure.

We now turn to the study of (3.10) .

Proposition 3.6. The natural action of G on itself by left (right) translations defines a left Poisson group action $G_{(R,-R)} \times G_{(R,R')} \to G_{(R,R')}$ (correspondingly, a right Poisson group action $G_{(R,R')} \times G_{R',-R')} \to$) $\to G_{(R,R')}$).

Let us prove e.g. the first statement. Let λ_x, ρ_x be the left (right) translation operators by an element $x \in G$ acting in $C^\infty(G)$. It suffices to check that

$$(3.14) \ \{ \varphi, \psi \}_{R,R'} \ (xy) = \{ \lambda_x \varphi, \lambda_x \psi \}_{R,R'} \ (y) + \{ \rho_y \varphi, \rho_y \psi \}_{R,-R} (x)$$

for any $\varphi, \psi \in C^\infty(G)$, $x, y \in G$. Obviously, one has from (3.13)

$$(3.15) \ \eta'_{R,R'} \ (xy) = \eta'_{R,R'} \ (y) + Ad_{y^{-1}} \circ \eta'_{R,-R} \ (x) \cdot Ad_y .$$

Clearly, (3.15) and (3.6) imply (3.14) .

Note . We may convert a right Poisson action into a left one by changing the sign of the Poisson bracket on G .

Proposition 3.6. provides us with the first example of Poisson group actions. Further examples will appear later on.

It is useful to have an infinitesimal characteristic of Poisson group actions. Let G be a connected Poisson Lie group, $(\mathcal{G}, \mathcal{G}^*)$ is tangent Lie bialgebra. Suppose that G is acting on a Poisson manifold $\mathcal{M}$ .For $\varphi \in C^\infty(\mathcal{M})$ let $\xi_\varphi (x) = d_g \varphi(g \cdot x)_{g=1}$ (where d_g means differential in the 1st variable $g \in G$). Let $\hat{X}$ be the vector field on $\mathcal{M}$ generated by an element $X \in$.

Proposition 3.7. The action of G is a Poisson group action if and only if

$$(3.16) \ \hat{X} \{ \varphi, \psi \}_{\mathcal{M}} - \{ \hat{X}_\varphi, \psi \}_{\mathcal{M}} - \{ \varphi, \hat{X}\psi \}_{\mathcal{M}} = \langle X, [\xi_\varphi, \xi_\psi]_* \rangle$$

for any $\varphi, \psi \in C^\infty(\mathcal{M})$, $X \in \mathcal{G}$.

The direct claim follows immediately from the definitions. Notice that if the right hand side is not zero identically in $\varphi, \psi \in C^\infty(\mathcal{M})$ the vector field $\hat{X}$ is certainly non-Hamiltonian.

3.4. The squares of Lie algebras.

Let $(\mathcal{G}, \mathcal{G}_R)$ be a Baxter Lie algebra. Recall from proposition 1.1 that $\mathcal{G}_R$ is embedded into $\mathcal{G} \oplus \mathcal{G}$ via (1.9). Let $\delta\mathcal{G} \subset \mathcal{G} \oplus \mathcal{G}$ be the diagonal subalgebra. We denote $d = \mathcal{G} \oplus \mathcal{G}$ and equip d with the inner product

$$(3.17) \quad \langle\!\langle (X_1, X_2), (Y_1, Y_2) \rangle\!\rangle = \langle X_1, Y_1 \rangle - \langle X_2, Y_2 \rangle.$$

Proposition 3.8. (i) There is a direct sum decomposition

$$(3.18) \quad d = \delta\mathcal{G} \dotplus \mathcal{G}_R.$$

(ii) The subspaces $\delta\mathcal{G}$, $\mathcal{G}_R \subset d$ are isotropic with respect to (3.17).

Let $P_{\delta\mathcal{G}}$, $P_{\mathcal{G}_R}$ be the projection operators onto $\mathcal{G}$, $\mathcal{G}_R$ in the decomposition (3.18). Put

$$(3.19) \quad R_d = P_{\delta\mathcal{G}} - R_{\mathcal{G}_R}.$$

Proposition 3.9. The operator (3.19) is skew with respect to (3.17) and satisfies the Yang-Baxter identity (1.8).

Hence R_d defines the structure of a Baxter-Lie algebra on d. We shall refer to (d, d_{R_d}) as the square of $(\mathcal{G}, \mathcal{G}_R)$.

Note. A similar construction works for arbitrary double Lie algebras (with the effect that R_d is no longer skew). This shows that by enlarging our Lie algebras we can always replace an arbitrary $R \in \mathrm{End}\ \mathcal{G}$ satisfying (1.8) with an operator of the form (2.2). We leave it to the reader to show that by applying Theorem 1.1. to Casimir functions on d^* restricted to $\mathcal{G}_R^* \subset (\delta\mathcal{G})^\perp \subset d^*$ we get the same Lax equations as in (1.6).

Proposition 3.9 is a special case of the following more general result which holds for arbitrary Lie bialgebras.

Theorem 3.2 /5/. (i) Let $(\mathcal{G}, \mathcal{G}^*)$ be a Lie bialgebra. There is a unique structure of a Lie algebra in the space $d = \mathcal{G} \dotplus \mathcal{G}^*$ such that (a) $\mathcal{G}$, $\mathcal{G}^* \subset d$ are Lie subalgebras; (b) The inner product on d given by $\langle\!\langle (X_1, f_1), (X_2, f_2) \rangle\!\rangle = f_1(X_2) + f_2(X_1)$ is invariant.
(ii) Conversely, suppose d is a Lie algebra equipped with an invariant inner product and $\mathcal{G}$, h are two its Lie subalgebras such that (a) $d = \mathcal{G} \dotplus h$ (b) $\mathcal{G}$, $h \subset d$ are isotropic subspaces. Then $(\mathcal{G}, h)$

is a Lie bialgebra. (iii) Natural embeddings $\mathcal{g}$, $*\subset$ d induce Lie
bialgebra morphisms $(\mathcal{g}$, $h^*)$ $\to$ (d, d_{-R_d}) , $(h, \mathcal{g})$ $\to$ (d, d_{R_d}).

Corollary. Suppose $(\mathcal{g}, \mathcal{g}^*)$ is a Lie bialgebra. Then $(\mathcal{g}^*, \mathcal{g})$ is
also a Lie bialgebra.

We shall refer to $(\mathcal{g}^*, \mathcal{g})$ as the dual Lie bialgebra of $(\mathcal{g}, \mathcal{g}^*)$. In
order to obtain the global counterpart of Theorem 3.2 let us quote
first the following corollary of Theorem 3.1.

Theorem 3.3. Let H be a Poisson Lie group and K $\subset$ H its subgroup.
Let k, f be the corresponding Lie algebras. Suppose that $k^\perp \subset h^*$ is
an ideal. Then K $\subset$ H is a Poisson submanifold and induced Poisson
bracket on K equips it with the structure of a Poisson Lie group. Its
tangent Lie bialgebra is $(k , h^*/k^\perp)$.

By applying Theorem 3.3 to our present situation we get the following
result.

Let $\mathcal{D}, G, G^*$ be the Poisson Lie groups whose tangent Lie bialgebras are
(d, d_{-R_d}) , $(\mathcal{g}, \mathcal{g}^*)$, $(\mathcal{g}^*, \mathcal{g})$ respectively.
Proposition 3.10. (i) Natural embeddings $G \to \mathcal{D}, G^* \subset \mathcal{D}$
are Poisson (anti-Poisson) mappings.

Note. Instead of using Theorem 3.1. to prove the above result we may
analyse the restriction of Poisson bracket to $G, G^* \subset \mathcal{D}$ directly.
This permits actually to prove Theorem 3.1. Indeed, on $\mathcal{D}$ the 1-cocy-
cle η is always trivial. By restricting it to $G \subset \mathcal{D}$ we get a (not
necessarily trivial) cocycle which determines the Poisson structure on
G (cf. Proposition 3.11 below).

Since $d = \mathcal{g} + \mathcal{g}^*$, $G . G^* \subset \mathcal{D}$ is an open cell Proposition 3.10
can be refined as follows.

Proposition 3.11. Change the sign of Poisson bracket on G^*. Then
the natural mapping $G \times G^* \to \mathcal{D}$: $(g, g^*) \mapsto gg^*$ is a Poisson
immersion.

We shall give a proof in §4 .

Examples 3.2. Poisson Lie groups described in Example 3.1. are dual
to each other. The Lie group which corresponds to $d = \mathcal{g} \oplus \mathcal{g}^*$ is
$T^*G \simeq G \times \mathcal{g}^*$ equipped with the Lie Poisson bracket depending on a

second argument.

We now return to the situation of Proposition 3.9 . Our next step will be to describe explicitly the Poisson structure on G_R which corresponds to the Lie bialgebra (a_R, a) . We shall need the following simple formula.

Lemma. Coadjoint action of the group G_R is given by

$$(3.20) \qquad Ad^*_{G_R} \, h \cdot X = (Ad_G \, h_- \, X)_+ - (Ad_G \, h_+ \cdot X)_- \,,$$

where $G_R \to G \qquad : h \mapsto h_\pm$ are natural homomorphisms and $X_\pm = 1/2(\, RX \pm X \,)$.

Proof. Decomposition (3.18) allows to identify a^*_R with δa. It is easy to check that $Ad^*_{G_R} \, h \cdot (X, X) = P_{\delta a} (Ad_G \, h_+ \cdot X, \, Ad_G h_- \cdot X)$ which yields (3.20) .

Proposition 3.12. The Poisson structure on G_R with the tangent Lie bialgebra (a_R, a) is given by

$$(3.21) \qquad \{ \varphi, \psi \} (h) = \langle\!\langle \eta (h) X, Y \rangle\!\rangle \,, \quad X = D\varphi(h), \, Y = D\psi(h)$$

where $\eta (h) \in Hom \, (\delta a, a_R)$ is given by

$$(3.22) \qquad \eta (h) \, X = (Ad_G \, h_+ \circ Ad^*_{G_R} \, h^{-1} \cdot X - X, \, Ad_G h \cdot Ad^*_{G_R} h^{-1} \cdot X - X).$$

The proof follows immediately from Proposition 3.10 (we restrict the 1-cocycle on $\mathcal{D} = G \times G$ to G_R and project it down to get a linear mapping acting from $\delta a \simeq a^*_R$ into a_R) . Note in particular that

$$(3.22) \qquad \eta (h_g) = Ad_{G_R} \, h \circ \eta (g) \circ Ad^*_{G_R} \, h^{-1} + \eta (h)$$

which implies that (3.22) defines indeed a Poisson Lie group structure on G_R .

§4. POISSON REDUCTION. DEFINITION AND APPLICATIONS

Let $\mathcal{M}$ be a Poisson manifold. Suppose that a Lie group G is acting on $\mathcal{M}$ and that the space of G -orbits is a smooth manifold.

Definition 4.1. An action $G \times \mathcal{M} \to \mathcal{M}$ is called admissible if the space of invariants $C^\infty(\mathcal{M})^G \subset C^\infty(\mathcal{M})$ is a Lie subalgebra.

If the action of G is admissible there is a unique Poisson structure on $G \backslash \mathcal{M}$ such that the natural projection $\pi: \mathcal{M} \to G \backslash \mathcal{M}$ is a Poisson mapping. We shall refer to $G \backslash \mathcal{M}$ as the reduced Poisson manifold. The concept of Poisson reduction extends the Poisson approach to the Hamiltonian reduction which was exposed in /16/ (and goes back to S.Lie). It was originally suggested by V.Drinfel'd.

Various examples of admissible group action are provided by the following result:

Proposition 4.1. Let G be a Poisson Lie group, $H \subset G$ its connected Lie subgroup, $h \subset \mathcal{G}$ the corresponding Lie algebra. Suppose that $h^{\perp} \subset \mathcal{G}^{*}$ is a Lie subalgebra. Let $G \times \mathcal{M} \to \mathcal{M}$ be a Poisson group action on a Poisson manifold $\mathcal{M}$. Then the space $C^{H} \subset C^{\infty}(\mathcal{M})$ of H-invariants is a Lie subalgebra, hence the restriction of our action to H is admissible.

In particular, the action of G itself is admissible (The point is that $H \subset G$ is not necessarily a Poisson Lie group).

Proof. We use Proposition 3.7. Let $\varphi, \psi \in C^{H}$. Then $\hat{X} \varphi = \hat{X} \psi = 0$ for $X \in h$ and $\xi_{\varphi}, \xi_{\psi} \in h^{\perp}$. By (3.16) we get

$$\hat{X} \{\varphi, \psi\} = \langle X, [\xi_{\varphi}, \xi_{\psi}]_{*} \rangle = 0$$

whence $\{\varphi, \psi\} \in C^{H}$.

As usual, the difficult part of reduction lies in the description of symplectic leaves in the quotient space. For Hamiltonian group actions this is done by means of the moment map. The point is to indicate its analog in the present case. We shall not do it in full generality (although the corresponding theorem was recently proved by Karasiov /17/) but rather present a series of examples.

Let $(\mathcal{G}, \mathcal{G}_{R})$ be a Baxter Lie algebra, $d = \mathcal{G} \oplus \mathcal{G}$ its square, $\mathcal{D} = G \times G$. Equip $\mathcal{D}$ with the Poisson bracket $\{,\}_{Rd, Rd}$ given by (3.12), (3.19).

Proposition 4.2. The Poisson bracket just described is non-degenerate on an open set in $\mathcal{D}$.

Proof. The operator $\gamma_{Rd, Rd}(x)$ defining the bracket is certainly non-degenerate at $x = $ identity and hence also on an open set in $\mathcal{D}$.

Note. The construction described extends straightforwardly to the square of arbitrary Poisson Lie group. In particular in the situation of examples 3.1, 3.2 we get the standard Poisson bracket on $\mathcal{D} \simeq T^{*}G$

Hence proposition 4.1 provides a generalization of cotangent bundles. This observation is due to V.Drinfel'd (unpublished).

From Propositions 3.6, 3.10, 4.1 we see readily that the action of G, G_R on D by left and right translations is admissible. Recall that ${}^sG$ G_R is open in $D = G \times G$. Hence we may identify $G(G_R)$ with an open cell in D/G_R (respectively, in D/G).

Proposition 4.3. (i) Natural projections

$$\pi \swarrow \overset{D}{} \searrow \pi \qquad\qquad s \swarrow \overset{D}{} \searrow s'$$
$$D/G_R \qquad G_R \backslash D \qquad\qquad D/G \qquad G/D$$

form dual pairs in the sense of /16/.

(ii) The quotient Poisson structure on $G \simeq D/G_R$ coincides with the Sklyanin bracket (3.12), (2.28) .

(iii) The quotient Poisson structure on $G_R \simeq D/{}^sG$ coincides with the bracket (3.21), (3.22) .

Proof. By definition, (i) means that right- G_R -invariant functions have zero Poisson brackets with left- G_R -invariant functions, and similarly for the second case. Suppose $\varphi \in C^\infty(D)$ is right-G_R invariant, $\psi \in C^\infty(D)$ is left- G_R-invariant. Then ∇'_φ , $\nabla_\psi \in a_R{}^\perp = a_R$. Hence

$$2 \{\varphi, \psi\}_{(Rd, Rd)} = \langle\langle R_d(\nabla'_\varphi, \nabla'_\psi)\rangle\rangle + \langle\langle R_d(\nabla_\varphi), \nabla_\psi\rangle\rangle =$$
$$= \langle\langle \nabla'_\varphi, \nabla'_\psi \rangle\rangle - \langle\langle \nabla_\varphi, \nabla_\psi\rangle\rangle = 0$$

The 2nd case is considered in a similar way.

We now pass to the calculation of the quotient Poisson structure on $G \simeq D/G_R$. Let $\varphi, \psi \in C^\infty(G)$, $X = \nabla_\varphi$, $X' = \nabla'_\varphi$, $Y = \nabla_\psi$, $Y' = \nabla'_\psi$. We extend φ, ψ to right- G_R -invariant functions $\hat{\varphi}, \hat{\psi}$ on $D\varphi$. Then

(4.1) $\nabla'_{\hat\varphi}\big|_{{}^sG \subset D} = (X'_+, X'_-)$, $\nabla'_{\hat\varphi}\big|_{x \in {}^sG} = (Ad_G x. X'_+, Ad_G x. X'_-)$

and similarly for ψ . By definition

(4.2) $2\{\varphi, \psi\}_{red.}(x) = 2\{\hat\varphi, \hat\psi\}_D(x) = \langle\langle R_d \nabla_{\hat\varphi}, \nabla_{\hat\psi}\rangle\rangle + \langle\langle R_d \nabla'_{\hat\varphi}, \nabla'_{\hat\psi}\rangle\rangle$

Since $\nabla'_{\hat\varphi} \in a_R$ and $a_R \subset d$ is isotropic the 2nd term vanishes. One checks easily that

(4.3) $R_d(\xi, \eta) = (R\xi - 2\eta_+, 2\xi_- - R\eta),$

By substituting (4.1), (4.3) into (4.2) and making use of the definition (3.17) we get after some easy computation

$$2\{\varphi, \psi\}_{red.} = \langle R X', Y' \rangle - \langle RX, Y \rangle$$

which coincides with (3.12) . We leave it to the reader to prove the
last assertion which is done similarly.

As a corollary of Proposition 4.3. we get

Theorem 4.1. (i) Natural action $\mathcal{D} \times \mathcal{D}/G_R \to \mathcal{D}/G_R$ is a Poisson group
action. (ii) Let us identify the quotient space with $G \subset G$.
Then this action is given by the formula

$$(4.4) \quad (x, y): \xi \longmapsto x \xi (\xi^{-1} x^{-1} y \xi)_+ = y \xi (\xi^{-1} x^{-1} y \xi)_-$$

In particular, the subgroup $G_R \subset \mathcal{D}$ is acting via

$$(4.5) \quad h: \xi \longmapsto h_\pm \xi (\xi^{-1} h_+^{-1} h_- \xi)_\pm$$

This action is a Poisson group action and its orbits coincide with the
symplectic leaves in G (equipped with the Sklyanin bracket).

We shall call (4.5) the dressing action. It may be regarded as an analogue of the the coadjoint action.

Proof. Since natural projections

$$\mathcal{D}/G_R \xleftarrow{\mathcal{D}} \searrow G_R \backslash \mathcal{D}$$

form a dual pair, we are in a position to apply a general theorem from
/16/. It asserts that if (π, π') is a dual pair of Poisson mappings,
then the symplectic leaves are obtained by blowing up points in the
double fibering (π, π') , i.e. they are the connected components of
$\pi'(\pi^{-1})(x))$. The projection map $\pi: \mathcal{D} \to G_R \backslash \mathcal{D} \simeq G$ is given by

$$\pi(u, v) = (uv^{-1})_+^{-1} u = (uv^{-1})_-^{-1} v,$$

whence

$$\pi^{-1}(\xi) = \{(h_+ \xi, h_- \xi) ; h \in G_R\}$$

This makes the last assertion obvious. (All the rest is perfectly
evident).

Note. The result we have quoted is a slightly refined version of a
theorem due to V.Drinfel'd.

In a dual fashion we may give a description of symplectic leaves in
$G_R \simeq \mathcal{D}/G$. Note first of all that G serves as another model for

the quotient spaces $\mathcal{D}/G$, $G\backslash\mathcal{D}$. Canonical projections are then given by

(4.6) $\quad \rho : (x,y) \to xy^{-1} \qquad , \quad \rho' : (x,y) \to y^{-1}x$.

Corollary 1. Symplectic leaves in G_R are mapped onto conjugacy classes in G under the canonical mapping $m : G_R \to G$.

Proof. Both groups are different models of the same quotient space.

Corollary 2. Casimir functions on G with respect to the Poisson structure described in Proposition 4.4 are precisely the central functions on G .

For completeness we give an explicit formula for this Poisson structure

Proposition 4.7. The quotient Poisson structure on G is given by

(4.7) $2\{\varphi,\psi\}_{red.} = \langle R(x), y'\rangle + \langle R(x'), y\rangle - \langle R(x), y\rangle - \langle R(x'), y'\rangle$
$\qquad\qquad\qquad + \langle x, y'\rangle - \langle x', y\rangle$

where $\quad x = \nabla_\varphi$, $x' = \nabla'_\psi$, $y = \nabla_\psi$, $y' = \nabla'_\psi$.

We leave the proof to the reader (cf. the proof of Proposition 4.3).

As another application of the reduction technique we give a proof of Proposition 3.11.

Proposition 4.8. Canonical projections

$$\mathcal{D}_{(R_d, -R_d)}$$
$$G_R\backslash\mathcal{D} \swarrow \qquad \downarrow \quad \mathcal{D}/{}^sG$$

form a dual pair.

The proof is the same as in Proposition 4.3 (Note the sign difference in the Poisson bracket on $\mathcal{D}$!)

Corollary. Equip $\mathcal{D}/{}^sG \times G_R\backslash\mathcal{D}$ with the product Poisson structure. Canonical embedding $\mathcal{D}_{(R_d, -R_d)} \to \mathcal{D}/G \times G_R\backslash\mathcal{D}$ is a Poisson mapping.

It is easy to check that the quotient Poisson structures on $\mathcal{D}/{}^sG, G_R\backslash\mathcal{D}$ are again given by (3.21) - (3.22), (2.28) - (3.12), respectively. Since ${}^sG \cdot G_R \subset \mathcal{D}$ is an open subset this finishes the proof of Proposition (3.11).

§5. LAX EQUATIONS ON POISSON LIE GROUPS: A GEOMETRIC THEORY

We start with the simplest theorem on the subject which will then be generalized to include Lax equations for lattice systems. Throughout this § we assume that G is a Poisson Lie group and that its tangent Lie bialgebra is a Baxter Lie algebra. Recall from the end of §4 that there are two different Poisson structures on G which are given by (3.12-), (4.7). This suggests that we may use them to construct integrable systems in almost the same way as in Theorems 1.1, 1.2. As we shall see now, this is indeed the case.

Denote by $I(G)$ the space of Casimir functions for the bracket. (4.7)

Theorem 5.1. (i) Casimir functions of the Poisson bracket (4.7) are in involution with respect to the Sklyanin bracket (3.12-).
(ii) Let $\varphi \in I(G)$. The equation of motion defined by φ with respect to the Sklyanin bracket has the Lax form

$$(5.1) \quad \frac{dL}{dt} = [L, M] \quad , \quad M = \frac{1}{2} R\left(\nabla \varphi(L) \right) \quad , \quad L \in G$$

(iii) Let $x_{\pm}(t)$ be the solutions to the factorization problem (1.11) with the left hand side given by

$$(5.2) \quad x(t) = \exp t \nabla_\varphi(L)$$

The integral curve of the equation (5.1) starting at $L \in G$ is given by

$$(5.3) \quad L(t) = x_{\pm}(t)^{-1} L \, x_{\pm}(t).$$

The proof is parallel to the proof of Theorem 1.2. Observe first of all that left and right gradients of a function $\varphi \in I(G)$ coincide. This makes (i), (ii) directly obvious from the definition of Sklyanin bracket.

Proposition 5.1. Let $\varsigma : \mathcal{D} \to G : (x,y) \mapsto xy^{-1}$ be the standard projection, $\varphi \in I(G)$, $h_\varphi = \varphi \cdot \varsigma$. The integral curves of the Hamiltonian h_φ on $\mathcal{D}$ (R_1, R_1) are given by

$$(5.4) \quad \left(x_o \, e^{tX'}, \, y_o \, e^{tX'} \right) \quad , \quad X' = \nabla_\varphi'(x \cdot y^{-1})$$

Proof. Recall that ς is included into a dual pair (4.6). Projections of the integral curve in $\mathcal{D}$ onto the quotient spaces $\mathcal{D}/^\delta G$, reduce to points since the reduced Hamiltonians are Casimir functions. Since h_φ is both right- and left-G_R-invariant we have

$\nabla_{h\psi}$, $\nabla'_{h\psi}$ $\in$ δ_{ay} and for any $\psi \in C^\infty(\mathfrak{D})$
$$\{ h_\psi, \psi \} = \ll \nabla'_{h\psi}, \nabla'\psi \gg.$$
Obviously, $\nabla'_{h\psi} = (X',X') \in d$ where $X' = \nabla'_\psi (xy^{-1})$ is time-independent. Now (5.3) follows immediately.

Consider the action $G_R \times \mathfrak{D} \to \mathfrak{D}$ defined by

(5.5) $\qquad h: (x,y) \longmapsto (h_+ x h_-^{-1}, h_+ y h_-^{-1})$

Notice that the subgroup $(G,e) \subset \mathfrak{D}$ is a cross section of (5.4) on an open cell in $\mathfrak{D}$. Hence we get a canonical projection

(5.6) $p: \mathfrak{D} \to G : (x,y) \longmapsto y_+^{-1} xy_-, \quad y = y_+ y_-^{-1}.$

whose fibers coincide with G_R -orbits in $\mathfrak{D}$.

Proposition 5.2. (i) The action (5.5) is admissible. (ii) The quotient Poisson space is canonically isomorphic to $G(-R, R)$.

We shall prove a more general statement below (Theorem 5.4).

To finish the proof of Theorem 5.1 observe that for $\psi \in I(G)$ $h_\psi = \psi \cdot p$ hence (5.4) defines a quotient flow on $G(-R,R)$ with Hamiltonian ψ Projecting the flow (5.4) down to G gives (5.3).

We shall indicate a generalization of Theorem 5.2 which is suited for the study of lattice systems. Recall from Proposition 3.4 that we may use more general Poisson brackets given by (3.10), with the left and right R-matrices not necessarily coinciding. This observation is used to twist the Poisson bracket on $\mathfrak{D}$.

Let τ be an automorphism of a Baxter Lie algebra $(\mathscr{G}, R)$ i.e. an orthogonal operator $\tau \in$ Aut $\mathscr{G}$ which commutes with R. It gives rise to an automorphism of G which we denote by $g \longmapsto {}^\tau g$. Define the twisted conjugation $G \times G \longmapsto G$ by

(5.7) $\qquad g: h \longmapsto h g\, {}^\tau h^{-1}.$

Let ${}^\tau I(G)$ be the space of smooth functions on G invariant with respect to twisted conjugations.

Theorem 5.3. (i) Functions $\psi \in {}^\tau I(G)$ are in involution with respect to the Sklyanin bracket on G . (ii) Equations of motion defined by a Hamiltonian $\psi \in {}^\tau I(G)$ have the following form

(5.8) $\qquad \dfrac{dL}{dt} = LA - BL \qquad , \quad L \in G,$

with $B = \dfrac{1}{2} R (\nabla_\psi (L)), A = \tau(B)$. (iii) Let $x_+(t)$, $x_-(t)$ be the solutions to the factorization problem (1.11) with the left hand side given by

(5.9) $\quad x(t) = \exp t \, \nabla_\varphi (L).$

The integral curve of equation (5.8) defined by $\varphi \in \overset{\tau}{I}(G)$ are given by

(5.10) $\quad L(t) = x_\pm (t)^{-1} \cdot L^\tau x_\pm (t).$

The proof is based on the use of a twisted Poisson structure on $\mathcal{D}$.

Extend $\mathcal{Y}$ to $d = \mathcal{Y} \oplus \mathcal{Y}$ by

(5.11) $\quad \hat{\tau} (X, Y) = (X, \tau Y)$

and put

(5.12) $\quad {}^\tau R_d = \hat{\tau} \circ R_d \circ \hat{\tau}^{-1}$

We also put $\quad {}^\tau G = \hat{\tau} ({}^s G) = \{ (x, {}^\tau x) ; x \in G \} \subset \mathcal{D}.$

Equip $\mathcal{D}$ with the Poisson bracket (3.10) with $R = {}^\tau R_d$, $R' = R_d$.

Proposition 5.3. (i) The natural action of ${}^\tau G$ on $\mathcal{D}_{({}^\tau R_d, R_d)}$ by left translations is a Poisson action. (ii) The natural action of G on $\mathcal{D}_{({}^\tau R_d, R_d)}$ by right translations is a right Poisson action.

This is a corollary of Proposition 3.6 since ${}^s G \subset \mathcal{D}_{(R_d, -R_d)}, {}^\tau G$ are Poisson subgroups.

Proposition 5.4. Canonical projections

$\quad \beta : \mathcal{D}_{({}^\tau R_d, R_d)} \rightarrow \mathcal{D}/{}^s G, \quad \beta' : \mathcal{D}_{({}^\tau R_d, R_d)} \rightarrow {}^\tau G \backslash \mathcal{D}$

are dual to each other.

Both quotient spaces are naturally modelled on G. Projections β, β' are given by

(5.13) $\beta : (x, y) \mapsto x y^{-1}$, $\quad \beta' : (x, y) \mapsto \tau^{-1} y^{-1} \cdot x$

Proposition 5.5. Symplectic leaves with respect to the quotient Poisson structure on G are orbits of twisted conjugations (5.7).

Proof. It suffices to compute $\beta (\beta'^{-1} (x))$

Clearly, $\quad \beta'^{-1} (x) = \{ (\tau^{-1} y \, x, y) ; y \in G \}, \quad \beta (\beta'^{-1} (x)) =$
$\quad = \{ \tau^{-1} y \, x \, y^{-1} ; y \in G \}.$

Corollary. Casimir functions of the quotient Poisson structure on are invariants of twisted conjugations.

A generalization of formula (4.7) for the quotient Poisson structure on G is given by

(5.14) $2 \{ \varphi, \psi \}_{red.} = \langle R(\tau X), Y' \rangle + \langle R(X'), \tau Y \rangle - \langle R(X), Y \rangle -$
$\quad - \langle R(X'), Y' \rangle + \langle \tau X, Y' \rangle - \langle X', \tau Y \rangle$

Now everything is ready for the proof of Theorem 5.3.

Proposition 5.6. Let $\varphi \in I(G)$, $h_\varphi = \varphi \cdot \mathcal{S}$. Integral curves of the Hamiltonian h_φ on $\mathcal{D}(^\tau R_d, R_d)$ are given by (5.4).

We leave the proof to the reader since it is completely parallel to that of Proposition 5.1.

Consider the action $G_R \times \mathcal{D} \to \mathcal{D}$ given by

(5.15) $\quad h: (x,y) \mapsto \left(h_+ \, x \, h_-^{-1}, \, {}^\tau h_+ \, y \, h_-^{-1} \right)$

Theorem 5.4. (i) The action (5.15) is admissible. (ii) The quotient Poisson bracket on $\mathcal{D}/G_R \simeq G$ coincides with the Sklyanin bracket.

Proof. To check (i) we use Proposition 4.1.
Observe first of all that by combining left and right translations we get a Poisson group action:

$$\mathcal{D}_{(^\tau R_d, -^\tau R_d)} \times \mathcal{D}_{(-R_d, R_d)} \times \mathcal{D}_{(^\tau R_d, R_d)} \to \mathcal{D}_{(^\tau R_d, R_d)} :$$
$$(u,v): w \mapsto u \, w \, v^{-1}.$$

We have changed the sign of the Poisson bracket on the second copy of $\mathcal{D}$ so as to consider left actions (More generally, if there are two commuting Poisson group actions $G \times M \to M$, $H \times M \to M$, their combination gives rise to a Poisson group action of $G \times H$ (which is equipped with the product structure).

Now, G_R is embedded into $\mathcal{D} \times \mathcal{D}$ via

$$h \mapsto \left(h_+, \, {}^\tau h_+, \, h_-, \, h_- \right)$$

Since the tangent Lie bialgebra of $\mathcal{D} \times \mathcal{D}$ is $(d \oplus d, \, d_{-^\tau R_d} \oplus d_{*R_d})$ our claim follows, by virtue of Proposition 4.1, from the following lemma.

Lemma 1. $\mathcal{G}_R^\perp \subset d \oplus d$ $\qquad$ is a Lie subalgebra in $d_{-^\tau R_d} \oplus d_{R_d}$

Proof of the lemma. An element $(X_1, X_2, Y_1, Y_2) \in d \oplus d$ annihilates $\mathcal{G}_R$ if and only if

$$R_- (X_1 - \tau^{-1} X_2) + R_+ (Y_1 - Y_2) = 0$$

Equivalently, $\qquad\qquad (\xi, \xi' \in \mathcal{G} , \, \eta, \eta' \in \mathcal{G}_R , \, \eta_- + \eta'_+ = 0)$

$$\mathcal{G}_R^\perp = \left\{ \left((\xi, \tau\xi) + (\eta_+, \tau\eta_-) , \, (\xi', \xi') + (\eta'_+, \eta'_-) \right) \right\},$$

Since there are natural Lie algebra embeddings $\mathcal{G} , \, \mathcal{G}_R \subset d_{R_d}$,

$\mathcal{G}, {}^\tau \mathcal{G}_R \subset d^\tau_{R_d}$ it suffices to check that

$$R_- \eta_1 + R_+ \eta'_1 = 0 , \qquad R_- \eta_2 + R_+ \eta'_2 = 0$$

implies

$$R_- ([\eta_1, \eta_2]_R) + R_+ (- [\eta'_1, \eta'_2]_R) = 0 .$$

The last assertion follows immediately from the Yang-Baxter identity.

Indeed, $R_-([\eta_1, \eta_2]_R) = [R_-\eta_1, R_-\eta_2] = [R_+\eta_1', R_+\eta_2'] =$
$$= R_+([\eta_1', \eta_2']_R).$$

We now come to the proof of the second assertion of Theorem 5.4. Observe that the subgroup $(G, e) \subset \mathcal{D}$ is again a cross section of the action (5.15) on an open cell in $\mathcal{D}$. The canonical projection $P: \mathcal{D} \to G$ is now given by

$$P: (x,y) \mapsto x^{-1} y_+^{-1} \, x y_-$$

For $\varphi, \psi \in C^\infty(G)$ put $H_\varphi = \varphi \circ P$, $H_\psi = \psi \circ P$. Put $X = \nabla_\varphi$, $X' = \nabla'_\varphi$, $Y = \nabla_\psi$, $Y' = \nabla'_\psi$. It is easy to compute the gradients of H_φ. Their restrictions to the surface $(G, e) \subset \mathcal{D}$ are given by

$$\nabla H_\varphi = (X, X'_+ - \tau X_-), \quad \nabla'_{H\varphi} = (X', X'_+ - \tau X_-).$$

Similar formulae hold for the gradients of H_ψ. Now

$$Rd\,(\nabla' H_\varphi) = (X', X' - X'_- + \tau X_-)$$
$$\tau\,Rd\,(\nabla H_\varphi) = (2\tau^{-1} X'_+ - X_+ - X_-, X'_+ - \tau X_-).$$

After substituting these expressions into the definition of $\{\varphi, \psi\}_{(\tau Rd, Rd)}$ we get after some remarkable cancellations

$$2\{\varphi, \psi\}\big|_{(G,e)} = (X'_+, Y') - (X', Y'_+) - (X_+, Y) + (X, Y_+) = 2\{\varphi, \psi\}_{Skl}.$$

Note: Unfortunately, I do not know how to extend to the present case the qualitative argument which we have used in the proof of Theorem 1.2. This argument is now replaced by a direct computation.

Let us now apply Theorem 5.3 to the difference Lax equations. Let $(\mathcal{G}, R)$ be a Baxter Lie algebra, G the corresponding Poisson Lie group. Put $\mathcal{G} = \overset{N}{\oplus} \mathcal{G}, G = G^N$. We shall regard elements of G as functions mapping $\mathbb{Z}/N\mathbb{Z}$ into G. Equip $\mathcal{G}$ with the natural inner product

(5.16) $\langle X, Y \rangle = \sum_n \langle X_n, Y_n \rangle$

and extend $R \in \text{End}\,\mathcal{G}$ to $\mathcal{G}$ by setting $(RX)_m = R(X_m)$. This makes $(\mathcal{G}, R)$ a Baxter Lie algebra. Equip G with the product Poisson structure. Clearly, G is a Poisson Lie group and its tangent Lie bialgebra is $(\mathcal{G}, \mathcal{G}_R)$. We shall denote elements of G by $L = (L_1, \ldots, L_N)$. Define the mappings $\psi_m, T: G \to G$ by

(5.17) $\psi_m(L) = \overset{\leftarrow}{\prod_{1 \le k \le m}} L_k$, $\quad T(L) = \overset{\leftarrow}{\prod_{1 \le k \le N}} L_k$

Functions ψ_m satisfy the linear difference system

(5.18) $\psi_m = L_m \psi_{m-1}$, $\quad \psi_0 = 1$

while T is the monodromy matrix associated with (5.18). Obviously, one has

Proposition 5.7. The monodromy map $T: G \to G$ is a Poisson mapping.

This property of the Sklyanin bracket has served as a motivation for the whole theory. The quantum version of this statement goes back to R.Baxter.

Let $\tau \in$ Aut $\mathcal{G}$ be the cyclic permutation

$$(5.19) \qquad \tau: \; (X_1 \ldots, \; X_N) \; \longmapsto \; (X_N, \; X_1, \; X_2 \ldots X_{N-1})$$

Clearly, the twisted conjugations $L \longmapsto g \, L^{\tau} g^{-1}$ coincide with the gauge transformations for (5.18) induced by left translations $\psi_m \to g_m \psi_m$ in its solution space. The operator (5.19) preserves the inner product (5.16) and commutes with R . Hence Theorem 5.3 applies to the present situation. The space is described by the following simple theorem.

Theorem 5.5. ("Floquet") (i) Two elements $L, L' \in \mathbf{G}$ lie on the same gauge orbit in $\mathbf{G}$ if and only if their monodromy matrices $T(L)$, $T'(L')$ are conjugate in G . (ii) The algebra $^{\tau}I(\mathbf{G})$ is generated by the functions $h_\varphi, \; L \to \varphi(T(L))$, $\varphi \in I(G)$.

As a corollary of Theorem 5.3 we get

Theorem 5.6. (i) Functions h_φ , $\varphi \in I(G)$ are in involution with respect to the Sklyanin bracket on G . (ii) The Hamiltonian equation of motion with the Hamiltonian h_φ is given by

$$(5.20) \qquad \frac{d L_m}{dt} = L_m \, M_{m-1} - M_m \, L_m , \qquad M_m = \frac{1}{2} R \left(Ad \, \psi_{m-1} \left(\nabla_\varphi (T(L)) \right) \right)$$

(iii) Let $(g_m)_\pm (t)$ be the solutions to the factorization problem (1.11) with the left hand side given by

$$(5.21) \qquad g_m(t) = Ad \, \psi_{m-1}^\circ \cdot \ell xp \; t \, \nabla_\varphi \left(T(L^\circ) \right)$$

The integral curve of (5.20) with the origin at $L^\circ = (L_1^\circ, \ldots L_N^\circ)$ is given by

$$(5.22) \qquad L_m(t) = (g_m)_\pm (t)^{-1} \, L_m^\circ \, (g_{m-1})_\pm (t)$$

Note. A completely different approach to the study of difference Lax equations was described by B.Kupershmidt /18/. These two approaches may be linked together by a discrete version of the Drinfel'd-Sokolov theory / 19/. However, a detailed analyses of this link goes beyond the scope of the present paper.

§6. DRESSING TRANSFORMATIONS.

In the present paragraph we return to the study of Lax equations on the line described in §2.5. Our notation will be close to that introduced there, the only difference being that we now drop out the periodicity condition. Thus let $(\mathcal{G}, \mathcal{G}_R)$ be a Baxter Lie algebra. Let G, G_R be the corresponding dual Poisson Lie groups.

1. Let $\mathbf{g} = C_0^\infty(\mathbb{R}, \mathcal{G})$, $\mathbf{g}_R = C_0^\infty(\mathbb{R}, \mathcal{G}_R)$. We define an inner product on $\mathbf{g}$ by

(6.1)
$$\langle X, Y \rangle = \int dx \, \langle X(x), Y(x) \rangle$$

Clearly, $(\mathbf{g}, \mathbf{g}_R)$ is again a Baxter Lie algebra. The dual space $\mathbf{g}_R^*$ is equipped with the Lie Poisson bracket, i.e.

(6.2)
$$\{\varphi, \psi\}_{\mathbf{g}_R}(L) = \int dx \, \langle L(x), [d\varphi(L), d\psi(L)]_R(x) \rangle$$

We shall consider only smooth functionals on $\mathbf{g}_R$ and identify $\mathbf{g}_R^*$ with $C^\infty(\mathbb{R}, \mathcal{G})$. An element $Y \in C(\mathbb{R}, \mathcal{G})$ defines a linear functional on $\mathbf{g}_R$ by (6.1) .

Let us associate with each function $L \in \mathbf{g}_R^*$ the linear differential equation

(6.3)
$$\partial_x \, \tau = L\tau$$

(The charge e will be henceforth chosen to be 1, cf. (2.24). Let Ψ_L be its fundamental solution i.e. a function with values in G satisfying (6.3). We shall normalize by the condition $\Psi_L(0) = 1$.

Now we are in a position to state one of our main results. Let $m : G_R \rightarrow G : h \longmapsto h_+ h_-^{-1}$ be the canonical mapping (cf. Proposition 1.2). Define the "dressed" potential L^h , $L \in \mathbf{g}_R^*$, $h \in G_R$ by the formula

(6.4)
$$L^h = \left(\Psi_L \, m(h) \, \Psi_L^{-1} \right)_\pm^{-1} \left(-\partial_x + L \right) \left(\Psi_L \, m(h) \, \Psi_L^{-1} \right)_\pm$$

Here $x_\pm$ is defined as in (1.11) and the choice of the sign is irrelevant.

Theorem 6.1. Formula (6.4) defines a right Poisson action $G_R \times \mathbf{g}_R^* \rightarrow \mathbf{g}_R^*$

Note. In typical applications $\mathcal{G}$ is a simple Lie algebra. Thus although as a Lie group G_R often splits (e.g. $G_R = G_+ \times G_-$ if R is given by (2.2)) it does not contain Poisson Lie subgroups (cf. Theorems 3.1, 3.3) . Hence in the statement of Theorem 6.1 the group G_R cannot be replaced by a smaller subgroup.

Before we start a proof of Theorem 6.1 let us discuss the definition

(6.4) and its versions. Notice first of all that (6.4) follows from
the dressing transformation for wave functions

(6.5) $\psi^h = \left(\psi \, m(h) \, \psi^{-1} \right)_{\pm}^{-1} \psi \, h_{\pm}$

Observe indeed that

$$\left(\psi \, m(h) \, \psi^{-1} \right)_{+}^{-1} \psi \, h_{+} = \left(\psi \, m(h) \, \psi^{-1} \right)_{-}^{-1} \psi \, h_{-}$$

is an identity, so the choice of sign is indeed irrelevant. Clearly,
$\psi(0) = 1 \Rightarrow \psi^h(0) = 1$ hence ψ^h may be regarded as the fundamental solutions to the system $\partial_x \psi^h = L^h \psi^h$ whence we get for L^h the desired
formula (6.4). This argument shows in particular that (6.5) determines
a well defined transformation on the phase space $\mathcal{L}_{\mathcal{R}}^*$. Starting from
the papers /6/, /7/ people usually consider a different definition

(6.6) $\psi^g = \left(\psi \, g^{-1} \, \psi^{-1} \right)_{+}^{-1} \psi$

Unlike (6.5), this formula does not preserve the normalization condition $\psi(0) = 1$ and hence defines an action on the space of all solutions
to equations (6.3) which is a principal fibering over $\mathcal{L}_{\mathcal{R}}^*$. Notice
that (6.6) differs from (6.5) only by an apparently inessential cons-
tant factor. However, this factor leads to a drastic change in the
composition law of dressing transformations. The Poisson properties
of transformations (6.6) are not as nice as those of (6.5)-(6.4).
(cf. Theorem 6.2 below).

6.2. We proceed now to the proof of Theorem 6.1.
Its main steps are as follows. Since (6.5) is more easy to deal with
than (6.4) we start with the description of the Poisson structure on
the space V_0 of all normalized wave functions. The Poisson structure
on the space V of all wave functions will also be of importance.
However, -(6.5) is still too complicated, so it is unreasonable to
prove Theorem 6.1 by a direct computation. Instead we shall use the
reduction technique and identify V with the quotient of a larger
Poisson space. This will enable us to understand the geometrical
meaning of (6.5) and make the composition law for dressing transformations completely evident.
Let $V = C^\infty(R, G)$, $V_0 = \left\{ \psi \in V ; \ \psi(0) = 1 \right\}$.
By solving (6.3) we get a mapping $\psi: \mathcal{L}_{\mathcal{R}}^* \longrightarrow V_0 : L \mapsto v_L$. Let us
define a Poisson structure on V_0 by demanding that ψ is a Poisson
mapping. To characterize this structure we compute Poisson brackets of
"cylindrical" functionals on V_0 depending on the value of ψ at one

particular point $x \in \mathbb{R}$ each. Let $\phi \in C^\infty (G)$. Put $\Phi_x [\psi] = \Phi (\psi(x))$ $\psi \in V_o$. Denote by $\nabla_\phi(v), \nabla'_\phi(v)$ the left and right gradients of ϕ evaluated at $v \in G$.

Proposition 6.1. Let $\Phi_x, \tilde{\Phi}_y$ be two cylindrical functionals , $|x| \leqslant |y|$ Then for sgn x = sgn y we have

$$(6.7) \quad 2 \{ \Phi_x \cdot \psi, \ \tilde{\Phi}_y \cdot \psi \}_{\mathfrak{g}_R^*} (L) = \langle R (\nabla'_\phi (\psi_L(z))), \ \nabla'_{\tilde{\phi}} (\psi_L(y)) \rangle -$$
$$- \langle R (\nabla_\phi (\psi_L(x))), \ Ad \psi_L(x) \cdot \nabla'_{\tilde{\phi}} (\psi_L(y)) \rangle.$$

For sgn x ≠ sgn y the Poisson bracket vanishes.

Proof. The computation we are going to outline is a version of a standard one commonly used in the inverse scattering method.

Lemma 1. The gradient of the functional $L \longmapsto \Phi_x (\psi_L)$ is given by

$$(6.8) \quad (grad \ \Phi_x \cdot \psi (L)) (z) = Ad \ \psi_L(z) \cdot \nabla'_\phi (\psi_L(z)) \cdot \Theta (z - \tilde{z})$$

where $\Theta(x)$ is a step function, $\Theta(x) = 0$, $x < 0$, $\Theta(x) = 1$, $x \geqslant 0$.

This follows from routine perturbation theory computation applied to (6.3) .

Lemma 2. The gradient (6.8) satisfies the differential equation
$$(6.9) \quad \partial_z X = [L, X] \quad , \quad o \leqslant z \leqslant x$$
with the boundary conditions
$$(6.10) \quad X(o) = \nabla'_\phi (\psi_L (z)) , \quad X(x) = \nabla_\phi (\psi_L(x))$$

Let us now use (6.9), (6.10) to compute the left hand side of (6.7). Obviously,

$$2 \{ \Phi_x \cdot \psi , \ \tilde{\Phi}_y \cdot \psi \}_{\mathfrak{g}_R^*} (L) = \int_0^x dz \langle L(z), [R X_1, X_2] + [X_1, R X_2] \rangle =$$
$$= \int_0^x dz (\langle [L, X_1], R X_2 \rangle - \langle [L, X_2], R X_1 \rangle) = \int_0^x dz \ \partial_z \langle X_1, R X_2 \rangle =$$
$$= \langle \nabla_\phi (\psi_L(z)), R (\nabla'_{\tilde{\phi}} (\psi_L(y))) \rangle - \langle \nabla'_\phi (\psi_L(x)), R (\nabla'_{\tilde{\phi}} (\psi_L(y))) \rangle$$

We denoted $X_1(z) = (grad \ \Phi_x \cdot \psi_{[L]})(z)$, $X_2(z) = (grad \ \tilde{\Phi}_y \cdot \psi_{[L]})(z)$ and assumed that $|x| \leqslant |y|$, sgn x = sgn y . If the signs of x, y are different the right hand side vanishes since X_1, X_2 have disjoint supports.

We now use formula (6.7) to define a Poisson bracket on $V \supset V_o$. Thus we put
$$(6.11) \quad 2 \{ \Phi_x, \tilde{\Phi}_y \} (\psi) = \langle R (\nabla'_\phi (\psi(x)), \ \nabla'_{\tilde{\phi}} (\psi(y)) \rangle - \langle R (\nabla_\phi (\psi(x)),$$
$$, \ Ad \ \psi(x) \cdot \nabla'_{\tilde{\phi}} (\psi(y)) \rangle$$

if $|x| \leqslant |y|$, sign x = sign y , and
$$\{ \Phi_x, \Phi_y \} (\psi) = 0$$
if sgn x ≠ sgn y.

Proposition 6.2. $V_0 \subset V$ is a Poisson submanifold.

Proof. It is sufficient to show that for any $\phi, \tilde{\phi} \in C^\infty(G)$, $\{\phi_0, \tilde{\phi}_x\}$ vanishes on V_0. Obviously, we have

$$2\{\phi_0, \tilde{\phi}_x\} = \langle R\nabla'_\phi(\psi(0)), \nabla'_{\tilde{\phi}}(\psi(x))\rangle - \langle R(\nabla_\phi(\psi(0))), Ad\,\psi(0)\cdot\nabla'_{\tilde{\phi}}(\psi(x))\rangle = 0$$

if $\psi(0) = 1$.

6.3. As our next step we shall realize V as a quotient space of a larger Poisson manifold. Put $W = C^\infty(R, D)$. We define a Poisson structure on W by setting

$$(6.12) \quad 2\{\phi_x, \tilde{\phi}_y\}(w) = \langle\!\langle R_d(\nabla\phi(w(x)), Ad\,w(x)\cdot\nabla'_{\tilde{\phi}}(w(y)))\rangle\!\rangle - \langle\!\langle R_d\nabla'_\phi(w(x)), \nabla'_{\tilde{\phi}}(w(y))\rangle\!\rangle$$

for any two cylindrical functionals $w \mapsto \phi(w(x))$, $w \mapsto \phi(w(y))$, $|x| \leqslant |y|$. If sgn $x \neq$ sgn y the Poisson bracket is set to be zero. We are making use of the notation introduced in §3.4 (Proposition 3.8, 3.9).

Let us define an action of $G_R = C^\infty(R, G_R)$ on W by left translations

$$(6.13) \quad h_W(x) = h(x)w(x), \quad h(x) \in G_R \subset D$$

Lemma 3. The action (6.13) is admissible.

Proof. We must show that functionals on W which are invariant under (6.13) form a Lie subalgebra with respect to (6.12). Observe that if a functional ϕ_x is left-G_R-invariant, then $\phi \in C^\infty(G)$ is left-G_R-invariant and hence $\nabla_\phi \in G_R^\perp = G_R$. Thus (6.12) simplifies to give

$$(6.14) \quad \{\phi_x, \tilde{\phi}_y\}(w) = -\langle\!\langle (R_d)_+(\nabla'_\phi(w(x))), \nabla'_{\tilde{\phi}}(w(y))\rangle\!\rangle$$

Since right gradients are invariant under left translations the r.h.s. is again a left-G_R-invariant functional. Q.E.D.

The diagonal embedding $V \hookrightarrow W : \psi \mapsto (\psi, \psi)$ allows to consider V as a subspace of W. Since $W \simeq G_R \cdot V$ (at least locally) we may identify the quotient $G_R\backslash W$ with V.

Lemma 4. The quotient Poisson structure on $V \simeq G_R\backslash W$ coincides with (6.11).

Proof. Let $\phi_x, \tilde{\phi}_y$ be two cylindrical functionals on V. Assume that $|x| \leqslant |y|$, sgn $x =$ sgn y. We extend them to left-G_R-invariant functionals $\hat{\phi}_x, \hat{\phi}_y$ on W. One checks easily that for $\psi \in V \subset W$

$$(6.15) \quad \nabla\hat{\phi}_x[\psi] = (X_+, X_-) \in d, \quad X = \nabla_\phi(\psi(x)), \quad X_\pm = \tfrac{1}{2}(R\pm1)X.$$

Hence the right gradient is given by

$$(6.16) \quad \nabla'_{\hat{\phi}_x}[\psi] = \left(Ad\,\psi(x)^{-1}\,X_+ \,,\, Ad\,\psi(x)^{-1}\,X_- \right)$$

Substituting (6.16) into (6.14) we get, after some remarkable cancellations, (6.11). For completeness we reproduce this calculation. We have

$$\{\hat{\phi}_x, \hat{\phi}_y\}(\psi) = -\lll \rho_{say}\left(\nabla^k_{\hat{\phi}}(\psi(x))\right),\; \nabla^k_{\hat{\phi}}(\psi(y)) \ggg =$$

$$= -\lll \left(\left(Ad\,\psi(x)^{-1}\,X_-\right)_+ - \left(Ad\,\psi(x)^{-1}\,X_+\right)_- ,\; \left(Ad\,\psi(x)^{-1}\,X_-\right)_+ - \left(Ad\,\psi(x)^{-1}\,X_+\right)_-\right),$$

$$,\; \left(Ad\,\psi(y)^{-1}\,Y_+ \,,\, Ad\,\psi(y)^{-1}\,Y_-\right) \ggg =$$

$$= \left\langle\left(Ad\,\psi(x)^{-1}\,X_-\right)_+ ,\; Ad\,\psi(y)^{-1}\,Y_+\right\rangle + \left\langle\left(Ad\,\psi(x)^{-1}\,X_+\right)_- ,\; Ad\,\psi(y)^{-1}\,Y_+\right\rangle +$$

$$+ \left\langle\left(Ad\,\psi(x)^{-1}\,X_-\right)_+ ,\; Ad\,\psi(y)^{-1}\,Y_-\right\rangle - \left\langle\left(Ad\,\psi(x)^{-1}\,X_+\right)_- ,\; Ad\,\psi(y)^{-1}\,Y_-\right\rangle =$$

$$= \left\langle Ad\,\psi(x)\,X_- \,,\, Y'_-\right\rangle - \left\langle Ad\,\psi(x)^{-1}\,X_+ \,,\, Y'_+\right\rangle =$$

$$= \left\langle X'_- , Y'_- \right\rangle - \left\langle X_- , Ad\,\psi(x)\,Y'\right\rangle$$

We denoted $X = \nabla_\phi(\psi(x))$, $X' = \nabla'_\phi(\psi(x))$, $Y = \nabla_{\tilde{\phi}}(\psi(y))$, $Y' = \nabla'_{\tilde{\phi}}(\psi(y))$ and make use of simple properties of homomorphisms $\xi \mapsto \xi_\pm = 1/2\,(R \pm 1)\,\xi$.

Define the "diagonal" right action $W \times \mathcal{D} \to W$ by setting

$$(6.17) \quad g:\quad w \mapsto w\,g \qquad,\quad g \in \mathcal{D}$$

Lemma 5. Equip $\mathcal{D}$ with the standard Poisson structure. Then (6.17) is a Poisson group action.

Proof. Let us rewrite (6.12) using left-invariant frame

$$2\{\hat{\phi}_x, \hat{\phi}_y\}(w) = \lll \eta_x\left(\nabla'_\phi(w(x))\right),\; \nabla'_{\tilde{\phi}}(w(y)) \ggg$$

where

$$\eta_x[w] = Ad\,w(x)^{-1}\cdot Rd \circ Ad\,w(x) - Rd$$

We have

$$\eta_x[wg] = Adg^{-1}\circ Rd \circ Adg + Adg^{-1}\circ Rd \cdot Adg - Rd$$

Since $\nabla'_\phi(w(x)g) = Adg^{-1}(\nabla'_{g\circ\phi}(w(x)))$, our assertion immediately follows from the definition of Poisson group actions.

By projecting down the action (6.17) to the quotient space $G_R \backslash W \simeq V$ we get the following transformation

$$(6.18) \quad (x, y):\quad \psi \mapsto \psi_\odot(x,y) = \left(\psi x y^{-1}\psi'^{-1}\right)_+^{-1}\,\psi x\,\left(\psi x y^{-1}\psi^{-1}\right)_-^{-1}\,\psi y$$

Theorem 6.2. Formula (6.18) defines a right Poisson group action $V \times \mathcal{D} \longrightarrow V$.

This is an immediate corollary of Lemma 5.

Indeed the actions (6.17), (6.18) are included into a commutative
diagram.

$$\begin{array}{ccc} W \times \mathcal{D} & \to & W \\ \downarrow & & \downarrow \\ V \times \mathcal{D} & \to & V \end{array}$$

We point out the following special cases

(i)   The action of $G_R \subset \mathcal{D}$ is a Poisson group action. This action preserves V_0 and its restriction to V_0 coincides with (6.5)

(ii)   The action of $^S G \subset \mathcal{D}$ is a Poisson group action. Clearly it is
given simply by $\Psi(x, x) = \Psi x$. The quotient space is isomorphic to
$\mathcal{G}_R^*$. The natural projection is given by $\pi: \psi \to \partial_x \psi\, \psi^{-1}$.

(iii)   The action of subgroup $(e, G) \subset \mathcal{D}$ on V is given by (6.6). Unlike the preceding cases, this is not a Poisson group action. Indeed,
$(0, \mathcal{G})^\perp = (\mathcal{G}, 0)$ is not even a Lie subalgebra in d_{R_d} .

Of course, formula (3.16) provides a precise information on the "non-
conservation" of Poisson brackets even in this case. Now, however, the
right hand side is not defined in intrinsic terms. It depends rather
on the Poisson structure on the larger group $\mathcal{D} \supset (G, e)$ or on its
tangent Lie bialgebra.

6.4.   Dressing transformations and dynamics.

For concreteness we shall assume throughout this nº that
$\mathcal{G} = \mathrm{sl}(2, \mathbb{C}[\lambda, \lambda^{-1}])$. The standard decomposition

(6.19) $\mathcal{G} = \mathcal{G}_+ + \mathcal{G}_-,\quad \mathcal{G}_+ = \mathit{sl}(2, \mathbb{C}[\lambda]),\quad \mathcal{G}_- = \lambda^{-1} \mathit{sl}(2, \mathbb{C}[\lambda^{-1}])$

gives rise to an r-matrix on $\mathcal{G}$

(6.20) $R = P_+ - P_-$

which is skew with respect to the inner product

(6.21) $\langle X, Y \rangle = \mathrm{Res}_{\lambda=0}\ \mathrm{tr}\ X(\lambda)\, Y(\lambda).$

In physical literature (6.20) is often referred to as the (classical)
Yang r-matrix.

Let G, G_+, G_- be the corresponding Lie groups (i.e. the loop group
of sl(2) and its subgroups consisting of absolutely convergent Laurent
series). The factorization problem associated with (6.20) is the
standard matrix Riemann problem. Let $h \subset \mathrm{sl}(2)$ be the standard Cartan

subalgebra, $H \subset G$ its centralizer in G, $H_\pm = H \cap G_\pm$. Let $\psi_0(x) = \exp \sigma x$, $\sigma = \mathrm{diag}(1,-1)$. Put

$$X_m = \lambda^n \delta \quad, \quad t = (t_1, \ldots, t_n, \ldots) \quad, \quad t.X = \sum_{n=1}^{\infty} t_n X_n.$$

$$(6.22) \quad \psi_0(t,x) = \psi_0(x) e^{tX} = \psi_0(x) \odot (e^{t.X}, e^{t.x})$$

Let $\psi \in V_0$ be a wave function obtained from ψ_0 by a dressing transformation (6.18). Since $G_+ \subset G_R$ centralizes ψ_0, we may assume without any loss of generality that

$$(6.23) \quad \psi = \psi_0 \odot (1, g_-) \quad, \quad g_- \in G_-$$

the element g_- being defined uniquely up to a right factor which belongs to H_-. Dressing transformation defined by g_- transforms the "free wave function" $\psi_0(x,t)$ into

$$(6.24) \quad \psi(x,t) = \psi_0(x,t) \odot (1, g_-) = \psi_0(x) \odot (1, e^{t.\mathrm{Ad}g_-^{-1}X}),$$

the element $\mathrm{ad}g_-^{-1}X$ being defined uniquely. By projecting (6.24) back to $V_0 \simeq V/\delta G$ we get

$$(6.25) \quad \tilde{\psi}(x,t) = \psi_0(1, (e^{t.\mathrm{Ad}g_-^{-1}\cdot X})_-)$$

Formula (6.25) defines an action of H_+ on the G_R-orbit of ψ_0.

Proposition 6.3. Vector fields on V_0 which correspond to this action are Hamiltonian.

By contrast, the apparently more simple action $V \times H_+ \to V$ given by (6.24) is __not__ Hamiltonian. This is easily checked using formula (3.16) to control the non-conservation of Poisson brackets under the transformations (6.24). This observation, however slightly puzzling, does not contradict of course to Proposition 6.3. Indeed, the point is that embedding $H_+ \hookrightarrow G_R$ given by

$$\exp t.X \longmapsto (\exp t.\mathrm{Ad}g_-^{-1}.X)_-$$

with some fixed $g_- \in G_-$ is __not__ a group homomorphism.

REFERENCES

1. Sklyanin E.K.   On complete integrability of the Landau-Lifshitz
   equation. Preprint LOMI. E-3-79, Leningrad: LOMI , 1980.

2. Belavin A.A., Drinfel'd V.G.   On the solutions of the classical
   Yang-Baxter equation. Funct. Anal. and its Appl.   , 16(1982),
   159-180.

3. Semenov-Tian-Chansky M.A. What is the classical r-matrix. Funct.
   Anal. and its Appl. 17(1983) , 259-272.

4. Kostant B. **Quantization** and representation theory. - In Proc.
   of the Research Symp. on Representations of **Lie groups**, Oxf.,
   1977, London Math. Soc.Lect. Notes Series, 1979, v.34.

5. Drinfel'd V.G.   Hamiltonian structures on Lie groups, Lie
   bialgebras and the geometrical meaning of Yang-Baxter equations.
   Sov. Math. Doklady 27 (1983), 68.

6. Zakharov V.E. , Shabat A.B.   Integration of nonlinear equations
   by the inverse scattering method. II. Funct. Anal. and its Appl.
   13 (1979) , 166-174.

7. Date E., Jimbo M., Kashiwara M., Miwa T.   Transformation groups
   for soliton equations. Proc.Japan. Acad. Sci. 57 A (1981),
   3806-3816; Physica 4D (1982), 343-365 . Publ.RIMS Kyoto Univ.
   18 (1982), 1077-1119.

8. Segal G., Wilson G.   Loop groups and the equations of KDV type,
   Publ. Math. I.H.E.S. 61 (1985), 4-64.

9. Wilson G.   Habillage et fonctions τ , C.R.Acad.Sci. Paris,
   299 (1984), 587-590.

10. Semenov-Tian-Chansky M.A.   Dressing transformations and Poisson
    group actions. Publ. RIMS Kyoto Univ. 21 (1986) , N6.

11. Reyman A.G., Semenov-Tian-Chansky M.A.   Reduction of Hamiltonian
    systems, affine Lie algebras and Lax equations. I, II . Invent.
    Math., 54 (1979), 81-100; 63(1981) , 423-432.

12. Adler M., Moerbeke P. Complete integrable systems, Euclidean Lie algebras and curves. Adv.Math., 38 (1980), 267-317.

13. Reyman A.G., Semenov-Tian-Chansky M.A. A new integrable case of the motion of the 4-dimensional rigid body. Comm. Math.Phys. (106, (1986), 161-172).

14. Cherednik I.V. Definition of τ-functions for generalized affine Lie algebras. Funct. Anal. Appl. 17 (1983), 243-244.

15. Reyman A.G., Semenov-Tian-Chansky M.A. Lie algebras and Lax equations with the spectral parameter on an elliptic curve, Zapiski Nauchn. Semin. LOMI (in Russian), v.155 (1986).

16. Weinstein A., Local structure of Poisson manifolds, J.Diff.Geom., 18 (1983), 523-558.

17. Karasjov M.V., To appear in Sov. Math. Izvestija (1986).

18. Kupershmidt B.A., Discrete Lax equations and differential-difference calculus, Asterisque 123 (1985).

19. Drinfel'd V.G., Sokolov V.V., Equations of KdV type and simple Lie algebras. Sov. Math. Doklady, 23 (1981), 457-462.

ON MONTE CARLO SIMULATIONS OF RANDOM LOOPS AND SURFACES

M.Karowski
Institut für Theorie der Elementarteilchen
Freie Universität Berlin
Arnimallee 14
D-1000 Berlin 33

1 Introduction

I would like to report on some results [1-5] obtained in collaboration with W.Helfrich, F.Rys, R.Schrader, and H.J.Thun at the "Freie Universität Berlin". Random loops and surfaces are useful concepts in different regions of physics.

Quantum field theories can be formulated in terms of random walks [6]. Symanzik's [7] polymer description of euclidian quantum field theories can be understood for the simple case of free bosons as follows. The "partition" function of the theory

$$Z = \int d\phi \, \exp\left[-\int dx \, \phi \, (-\Box + m^2) \, \phi\right] \tag{1}$$

reads in regularized form on a lattice L^d

$$Z = \int d\phi \, \exp\left[-\phi^T \, (1 - \Gamma) \, \phi\right] \tag{2}$$

where ϕ is a L^d-dimensional vector and the matrix Γ is given by

$$\Gamma_{xy} = k \sum_{\mu=1}^{d} \left(\delta_{x+e_\mu, y} + \delta_{x-e_\mu, y} \right) \quad , \quad k = \frac{1}{2d + m^2} \quad . \tag{3}$$

The integrations in eq.(2) can be performed

$$Z = 1/ \det (1 - T)$$
$$= \exp \left[- \operatorname{tr} \ln (1 - T) \right] \qquad (4)$$
$$= \exp \sum_{\ell=1}^{\infty} \frac{1}{\ell} \operatorname{tr} T^{\ell} \; .$$

Since the matrix T "connects" nearest neighbours on the lattice, each term $\frac{1}{\ell} \operatorname{tr} T^{\ell}$ in eq.(4) represents a sum over all loops of length ℓ consisting of simple bonds. Thus the partition function can be written in terms of a statistical system

$$Z = \exp \sum_{conf.} k^{\ell} \qquad (5)$$

where the sum extends over all (possibly overlapping) one-loop configurations. The configurational energy $-T \cdot \ln k \cdot \ell$ is proportional to the total length of the loop. Expanding the exponentional in eq. (5) we get a representation analogous to the well known high temperature expansion for classical spin systems

$$Z = \sum_{\{s\}} \exp k \sum_{\langle xy \rangle} s_x s_y$$
$$= \sum_{\{s\}} \sum_{\ell=0}^{\infty} \frac{1}{\ell!} \left(s^T T s \right)^{\ell} \qquad (6)$$
$$= \sum_{conf.} c_{\ell} k^{\ell}$$

where a configuration contributing to the sum now may consist of several disconnected loops. The entropy factors c_{ℓ} depend on the model. For the Ising model one obtains after resummation a sum over nonoverlapping loops

$$Z = \sum_{conf.} e^{-\beta \ell} \quad , \quad e^{-\beta} = \operatorname{th} k \; . \qquad (7)$$

The O(N) nonlinear σ-model in the limit N→0 describes self-avoiding random walks [8,9].

The polymer formulation of fermions leads (due to Pauli's princip-

le) also to self-avoiding loops. This can be seen as follows [1].
The "partition function" for free fermions is

$$Z_F = \int d\psi^+ d\psi \, \exp\left[-\int dx \, \psi^+ (\gamma\partial + m)\psi\right]$$

$$= \det (\gamma\partial + m)$$

$$\propto \det (1 + \Gamma) \tag{8}$$

$$= \sum_{\pi} sgn \prod_x |(1 + \Gamma)_{x \, \pi(x)}|$$

where $sgn = sgn(\pi) \cdot sgn(1+\Gamma)_{x \, \pi(x)}$ and

$$\Gamma_{xy} = k \sum_{\mu=1}^{d} \gamma_\mu(x) (\delta_{x+e_\mu, y} - \delta_{x-e_\mu, y}), \quad k = \frac{1}{2m}. \tag{9}$$

(Note that for Kogut-Susskind fermions γ_μ can be taken as an x-
dependent number.) Any permutation π is a product of cyclic per-
mutations $\pi_1, \ldots, \pi_s$ which can be represented as s oriented non-
intersecting loops. The fermion partition function can be written
as

$$Z_F = \sum_{conf.} sgn \, k^\ell \tag{10}$$

where the sum extends over all configurations of oriented self-
avoiding loops of total length ℓ. Introducing an auxiliary statistical system by

$$Z^{polymer} = \sum_{conf.} k^\ell \tag{11}$$

we can write the fermion expectation value of an observable X as

$$\langle X \rangle_F = \frac{\langle X \cdot sgn \rangle^{polymer}}{\langle sgn \rangle^{polymer}}. \tag{12}$$

Whereas normal quantum fields (defined on points) are related to random walks, one can describe gauge field theories in terms of random
surfaces. Wilson's [10] lattice gauge theory formulation is given by

$$Z = \int d\mathcal{U} \, \exp \frac{1}{g^2} \sum_{plaq.} tr (\mathcal{U}_1 \mathcal{U}_2 \mathcal{U}_3 \mathcal{U}_4) \tag{13}$$

where $U_1, \ldots, U_4 \in G$ (eg. G = SU(N)) are defined on the links around a plaquette. The high temperature expansion $(g \to \infty)$ of this expression leads analogously to eq.(6) to

$$Z = \sum_{conf.} c_s \left(\frac{1}{g^2} \right)^s \tag{14}$$

where the sum extends over all two dimensional closed surfaces of total area s. The entropy factors c_s depend on the model. In the limit $N \to 0$ [9,11] one gets

$$Z = \sum_{conf.} \left(\frac{N}{g^2} \right)^s N^{\chi} \tag{15}$$

where the sum is now restricted to self-avoiding surfaces and χ is the Euler characteristic.

The string quantization problem has been formulated as summation over random surfaces [12] . Similar as fermionic particles are represented by self-avoiding walks, fermionic strings may be related to self-avoiding surfaces. An investigation of the critical behaviour of random surfaces should lead to a better understanding of gauge and string theories in their continuum limit.

Moreover, there are posssible applications of random walks and surfaces to statistical and solid state physics in two and three dimensions. Self-avoiding random chains play an important role in polymer physics [13] . They have been studied in the context of defect-line mediated phase transitions (e.g. in liquid crystals) [14] and the equilibrium polymerisation of sulphur [15] . Self-avoiding surfaces in three dimensions might [16] be useful for the understanding of microemulsions whose stability relies on a balance of entropy and energy of their interfaces [17] . They can also be expected to describe properties of flexible two dimensional sheet polymers [13] . The self-avoiding random surface model is a natural generalization of the solid-on-solid model [18] which is useful to describe the roughening of crystal surfaces and interfaces.

In ref.[1] we developed a Monte Carlo simulation method for quantum field theories involving fermions on the basis of the polymer formulation eqs.(11,12). A statistical model of loops in d=2,3, and 4 dimensions (loop gas model) was proposed in [2] to study the influence of the excluded volume repulsion on the critical equilibrium properties of statistical line systems. The critical exponents α, β, γ, and δ were evaluated by means of the "critical window" method in ref.[4] for the loop gas model in two dimensions. In ref.[3] we explored the phase diagrams of self-avoiding random surface models in three and four dimensions with surface tension and curvature energies. The critical exponents α, β, γ, and δ for intersecting surface gas models in three dimensions were evaluated in ref.[5]. For other investigations and Monte Carlo simulations of random walks and surfaces see references in [1-5].

2 Models

The models to be considered are defined on square, cubic, hyper cubic lattices L^d with periodic boundary conditions in d=2,3,4-dimensions with $|L^d| \lesssim 10^4$. The partition functions are defined by

$$Z = \sum_{c \epsilon \ell} exp \left[- E_{(c)} / kT \right] \tag{16}$$

where the sets of configurations ℓ and the configurational energies E(c) depend on the specific model. We distinguish two types of configurations of lines (two dimensional surfaces):

$$\ell_i = \{ \text{closed intersecting loops (surfaces)} \} \tag{17}$$

$$\ell_{sa} = \{ \text{closed self-avoiding loops (surfaces)} \} . \tag{18}$$

A configuration $c \epsilon \ell_i$ comprises a collection of links (plaquettes) in the lattice such that each vertex (link) in c is contained in two or four links (plaquettes). For $c \epsilon \ell_{sa}$ each vertex (link) in c is contained in two links (plaquettes). Thus the lines (surfaces) may not intersect at a common vertex (link). But two locally distinct surfaces are allowed to touch at a vertex, they will be considered as disconnected this point. The energies may include three terms

$$E / kT = \beta \cdot \ell + \beta_i \cdot i + \beta_n \cdot n \tag{19}$$

for loops and

$$E / kT = \beta \cdot s + \beta_\ell \cdot \ell + \beta_x \cdot x \tag{20}$$

for surfaces. The first, the tension term is proportional to the total loop length ℓ (surface area s). The intersection energy is proportional to the number of intersection points i (links ℓ). The third contributions are curvature energies. They can also be understood as chemical potential terms of the topological quantities: number of loops n and Euler characteristic χ, respectively. We are interested in the nature of phase transitions (first or second order), phase diagrams and critical exponents of these models.

3 Monte Carlo Method

For simplicity I shall describe the method in terms of random loops. A configuration (c.f.eqs.(17,18)) can be generated on a computer iteratively as follows. Starting from an old configuration one gets a new one by a local change in a unit square. This means the replacement of empty links by occupied ones and vice versa. The four possible types of such changes are depicted in fig.1. For the self-avoiding case (17) one has to make sure that no crossings appear. By a Monte Carlo simulation one generates samples of equilibrium ensembles of configurations c_i. In the heat bath updating procedure we sequenti-

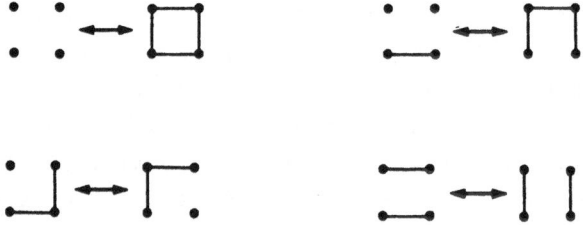

Figure 1. Local changes of loop configurations within a plaquette

ally sweep all d(d-1)/2 L^d plaquettes of the lattice and accept the new configuration with probability

$$P = w_{new}/(w_{old} + w_{new}) \tag{21}$$

where the w are the Boltzmann factors

$$\exp(-E/kT). \tag{22}$$

This means we take the new configuration if a pseudo-random number (equally distributed in the unit interval) is less than P, otherwise we retain the old one. Since every allowed configuration can in principle be attained after sufficiently many iterations the "ergodic condition" is satisfied. Obviously a large set of configurations with a probability distribution proportional to the Boltzmann factor (22) is stable under this procedure. Moreover, starting from an arbitrary initial configuration, we expect to reach such an equilibrium set after an appropriate "warming up" period. The thermal average of a variable A is calculated as the mean over $N \approx 10^3$ configurations each obtained after about five complete sweeps through the lattice

$$\langle A \rangle \approx \frac{1}{N} \sum_{i=1}^{N} A(c_i) . \tag{23}$$

The computations are usually done in "thermal cycles" T_{min}, $T_{min}+\Delta T$, ..., T_{max}, $T_{max}-\Delta T$, ..., T_{min} where we start at low temperature from the empty lattice.

4 Some Results

A) Self-avoiding loop gas

In ref.[2] we considered self-avoiding loop gas systems in d=2,3, and 4 dimensions

$$Z = \sum_{c \in \ell_{sa}} e^{-\beta \ell} \tag{24}$$

We "measured" the average length $\langle \ell \rangle$ (proportional to the energy) and the fluctuations $\langle \ell^2 \rangle - \langle \ell \rangle^2$ (proportional to the specific heat) in thermal cycles. For low temperatures a typical configuration consists of a few small loops. Around $\beta \approx 0.86$ $\langle \ell \rangle$ grows rapidly with temperature and $\langle \ell^2 \rangle - \langle \ell \rangle^2$ has a maximum. The configurations show strong fluctuations: some of them contain loops of medium size, others also comprise rather large ones. For high temperature the energy approaches its asymptotic value and the specific heat decreases again. The lattice is filled with many loops, among them a very large one. The characteristic dependence on β of $\langle \ell \rangle$ and $\langle \ell^2 \rangle - \langle \ell \rangle^2$ is displayed in fig.2. With increasing lattice size the peaks of the specific heat are higher and sharper.

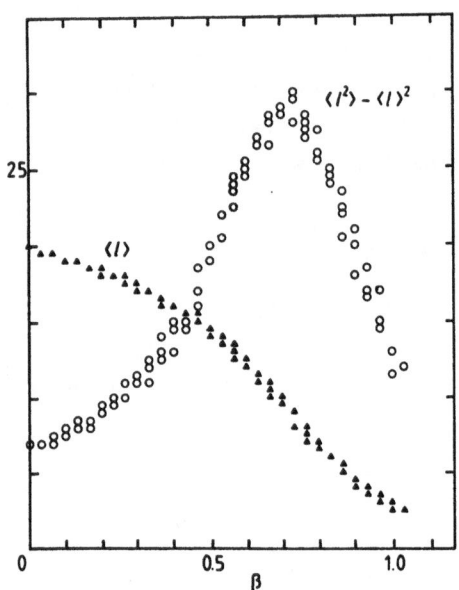

Figure 2. The average energy $\langle \ell \rangle$ and the length fluctuations $\langle \ell^2 \rangle - \langle \ell \rangle^2$ for a loop gas on a 5 x 5 lattice.

For $L \rightarrow \infty$ they approach

$$\langle \ell^2 \rangle - \langle \ell \rangle^2 \sim |\beta - \beta_{crit}|^{-\alpha} \tag{25}$$

a behaviour typical for a second order phase transition. We found
the critical points:

$$\beta_{crit} = \begin{cases} 0.86 & \text{in 2-dimensions} \\ 1.50 & \text{in 3-dimensions} \\ 1.9 & \text{in 4-dimensions.} \end{cases} \tag{26}$$

In ref.[4] we used the fact that in two dimensions loops divide the
lattice into "interior" and "exterior" parts, V_{in} and V_{ex}, respectively, to introduce an order parameter

$$m = \frac{V_{ex} - V_{in}}{V_{ex} + V_{in}} . \tag{27}$$

The critical exponents are defined by eq.(25) and

$$\left. \begin{array}{l} \langle m \rangle \sim |\beta - \beta_{crit}|^{\beta} \\ \langle \chi \rangle \sim |\beta - \beta_{crit}|^{-\gamma} \end{array} \right\} \quad \beta \to \beta_{crit} \,, \ h = 0$$

$$\langle m \rangle \sim |h|^{1/\delta} \qquad , \quad \beta = \beta_{crit} \,, \ h \to 0 \,. \tag{28}$$

χ is the susceptibility given by

$$\langle \chi \rangle \propto \frac{d}{dh} \langle m \rangle = \langle m^2 \rangle - \langle m \rangle^2 \tag{29}$$

and h a magnetic field introduced by the replacement $\beta\ell \to \beta\ell - hm$ in
eq.(24).

We were able to find a temperature regime within the critical region,
which is limited, close to T_{crit}, by the finite-size rounding temperature (where diverging thermodynamic quantities round off) on one
side, and the end of the critical regime away from T_{crit} (where cor-
rection-to-scaling terms become important) on the other side. In this
"critical window" (whose extension increases with increasing system
size) the critical exponents α, β, and γ where calculated from the
slope of the linear portion of the corresponding quantity in a doubly
logarithmical plot near T_{crit}. Furthermore, the exponent δ was determined from the isotherm at the critical temperature. We found
Ising-like values for all exponents considered.

B) Self-avoiding surfaces

In ref.[3] we investigated self-avoiding surface gas systems

$$Z = \sum_{c \cdot \ell_{sa}} exp\left[- \beta s + \mu x\right] \, .$$

(30)

The Euler characteristic is defined by

$$\chi = 2 \left(n_{comp} - n_{hand} \right)$$

(31)

where n_{comp} (n_{hand}) is the number of connected components (hand-les) of the surface. Fig.3 shows the average energy $\langle s \rangle$ and the

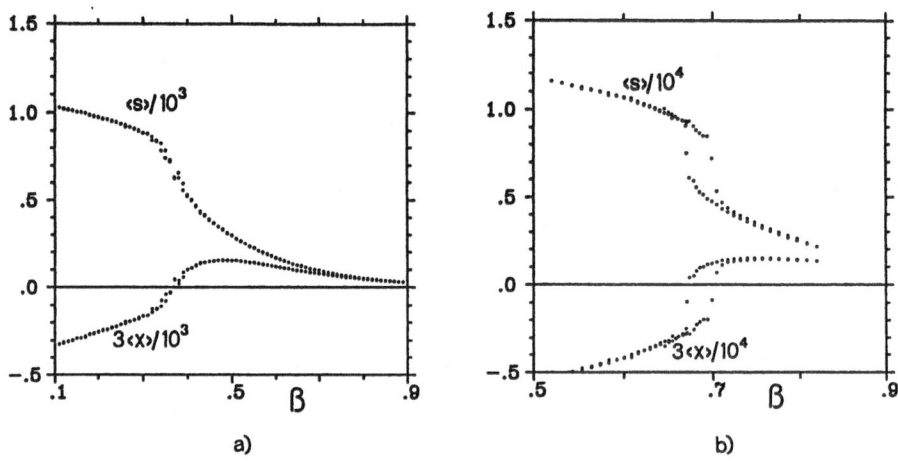

a)   b)

Figure 3. Monte Carlo results for the self-avoiding surface gas model. The average surface $\langle s \rangle$ and Euler characteristic were obtained on (a) 10^3 and (b) 10^4 lattices.

Euler characteristic obtained in "β-cycles" by Monte Carlo simu-lations for the case μ =0. In the low-temperature phase a typical configuration consists predominantly of small separated components. In the high-temperature phase χ is negative and a typical configuration consists of a single connected object with many handles like a sponge. In three dimensions (Fig.3a) the data show evidence of a second-order phase transition at

$$\beta_{crit} = 0.353 \quad (in\ 3\text{-dimensions}). \tag{32}$$

Note that the Euler characteristic χ vanishes at the critical point. This flatness might be related to scale invariance. In four dimensions (Fig.3b) we found hysterisis loops in the β-cycles indicating a first order phase transition at

$$\beta_{tran} = 0.68 \quad (in\ 4\text{-dimensions}) \tag{33}$$

with an energy jump $\Delta < s > L^{-4} = 0.45$. The observed different types of transitions are in agreement with those of lattice gauge theories in three and four dimensions.

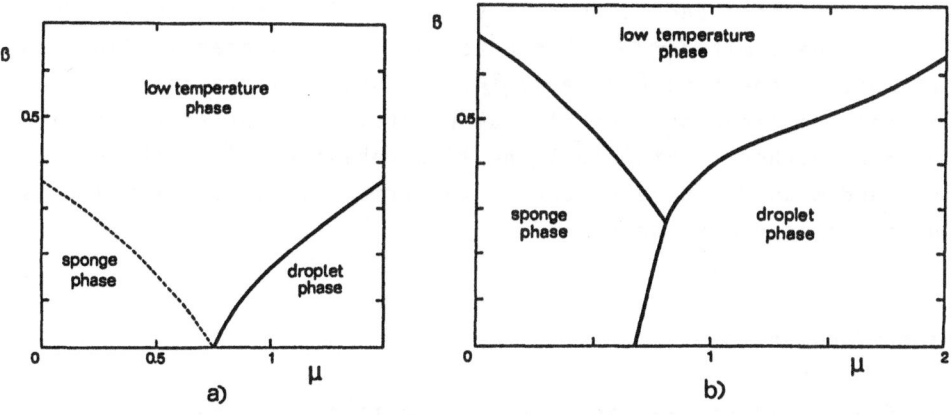

Figure 4. Phase diagrams for model (30) in (a) three and (b) four dimensions showing three phases, first (———) and second- (----) order transition lines.

For nonvanishing chemical potential for the Euler characteristic $\mu > 0$ we found the phase diagrams depicted in fig.4. For large μ a new phase appears separated from the others by first-order transitions. This "droplet phase" is related to a new ground state consisting of simple cubes, each touching eight others at its corners.

In ref.[5] the critical behaviour of the self-avoiding surface gas ($\mu = 0$) in three dimensions was investigated by the critical window method. Analogously to the loop gas case, I found Ising exponents $\alpha, \beta, \gamma,$ and δ.

C) Intersecting surfaces

An intersecting surface gas model in three dimensions

$$Z = \sum_{c \in \mathcal{C}_i} exp \left[-\beta \cdot s - \beta_\ell \cdot \ell \right] \tag{34}$$

previously discussed in [18] was investigated in ref. [5]. For $\beta_\ell = 0$ it is the Ising model and it approaches the self-avoiding model (30) (for $\mu = 0$) in the limit $\beta_\ell \to \infty$. The phase diagram depicted in fig.5 is symmetric with respect to $\beta \to -\beta - 2\beta_\ell$. It shows first- and second order transition lines and tricritical points. By means of the "critical window" method I obtained Ising-like behaviour along the critical lines and mean field behaviour at the tricritical points which is expected in three dimensions.

D) Hausdorff dimension

A model of a single self-avoiding random surface in three dimensions with the fixed topology of the sphere was considered in ref.[5]

$$Z = \sum e^{-\beta s} . \tag{35}$$

At the critical point $\beta_{crit} = 0.53$ [19] the average radius of gyration diverges like

$$\xi \sim |\beta - \beta_{crit}|^{-\nu} . \tag{36}$$

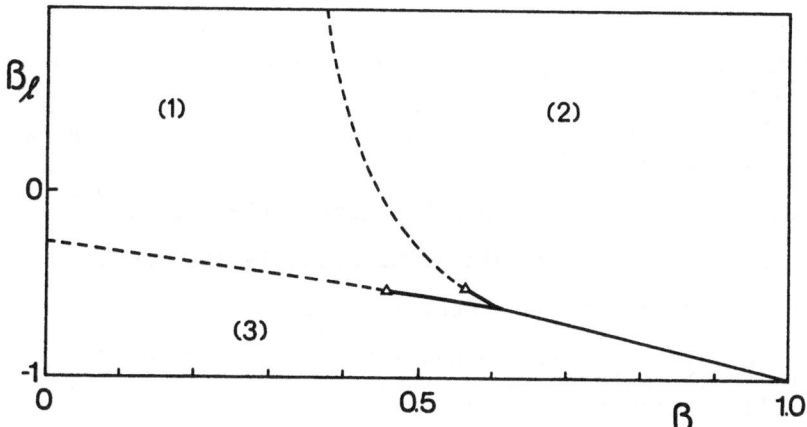

Figure 5. Phase diagram of the intersecting surface gas model showing a disordered (1), a ferromagnetic (2) and an antiferro magnetic phase (3) separated by first- (———) and second-order (-----) transition lines. At their juncture are tricritical points (△).

The critical exponent ν is related to the "Hausdorff dimension" of the surface

$$d_H = 1/\nu \qquad (37)$$

defined at $\beta = \beta_{crit}$ by

$$d_H = \lim_{R \to \infty} \langle \ln s(R) \rangle / \ln R \qquad (38)$$

where $s(R)$ is the part of the surface contained in a sphere with radius R, such that the surface passes is centre. The Monte Carlo result is

$$d_H = 2.30 \pm 0.05 \qquad (39)$$

in good agreement with a Flory-type formula $d_H = 2 \, ^1/3$ derived in [20].

References

1. M.Karowski, R.Schrader, and H.J.Thun, Commun.Math.Phys.97 (1985)5

2. M.Karowski, H.J.Thun, W.Helfrich, and F.Rys, J.Phys.A:Math. Gen.16(1983)4073

3. M.Karowski and H.J.Thun, Phys.Rev.Lett.54(1985)2556

4. M.Karowski and F.Rys, J.Phys.A: Math.Gen.19(1986)2599

5. M.Karowski, J.Phys.A: Math.Gen.(in press)

6. J.Fröhlich, in 'Progress in Gauge Field Theory', ed.G. t'Hooft et al (NATO Advanced Study Institute Series B No 115) (Plenum, New York 1984)

7. K.Symanzik, in 'Local Quantum Theory', Proc.Int.School of Physics 'Enrico Fermi', Course XLV, ed.R.Jost (Academic, New York 1969), p.152

8. P.de Gennes, Phys.Lett.38A(1972)339

9. A. Maritan and C.Omero, Phys.Lett.109B(1982)51

10. K.Wilson, Phys.Rev.D10(1974)2445

11. B.Durhuus, J.Fröhlich, and T.Jonsson, Nucl.Phys.B225 FS9 (1983)185

12. A.Polyakov, Phys.Lett.103B(1981)207

13. P.J.Flory, 'Principles of Polymer Chemistry' (Ithaca, N.Y.: Cornell University Press,1969)

14. F.Rys and W.Helfrich, J.Phys.A15(1982)599

15. J.C.Wheeler, S.J.Kennedy, and P.Pfeuty, Phys.Rev.Lett.45 (1980)1748

16. T.Hofsäss and H.Kleinert, Phys.Lett.102A(1984)420

17. P.de Gennes and C.Taupin, J.Phys.Chem.86(1982)2294

18. J.D.Weeks, 'Ordering in Strongly Fluctuation Condensed Matter Systems' ed T.Riste (Plenum, New York 1979)

19. T.Stirling and J.Greensite, Phys.Lett.121B(1983)345

20. A.Maritan and A.Stella, Phys.Rev.Lett.53(1984)123

FIELD THEORETIC METHODS IN CRITICAL PHENOMENA

WITH BOUNDARIES

A.M. Nemirovsky

The James Franck Institute

The University of Chicago, Chicago, IL 60637

ABSTRACT

Recent work on field theoretic methods in critical phenomena with boundaries by the author and colla-borators is described. The presence of interfaces and boundaries in critical systems produce a much richer set of phenomena than that of infinite sized systems. New universality classes are present and interesting crossover behavior occurs when there is a relative variation of additional length scales associated with either the size of the system or the boundary conditions (BC) satisfied by the order parameter on the limiting surfaces. A recently proprosed crossover renormalization group approach is very well suited to study these rich crossovers. Since functional integrals provide an indefinite integral representation of field theories, Feynman rules in configuration space are independent of geometry and BC. Renormalization of field theories with boundaries is discussed and various geometries and BC are considered. Application of field theoretic techniques are described for studying conformational properties of long polymer chains in dilute solution near interfaces or in confined domains. Also, related problems in quantum field theories with boundaries are presented.

The work I present here was performed in collaboration with K.F. Freed. Also, Z-G. Wang and J.F. Douglas have contributed to some of the work described below.

1. INTRODUCTION

Experiments and computer simulations can only probe finite systems with limiting surfaces. On the other hand, theoretical studies of phase transitions (PT) usually consider infinitely extended systems. Although surface effects can, in general, be neglected in large systems, these effects become very relevant near a second order PT point as the correlation length grows unbounded.[1] Critical singularities at a second order PT only occur in the thermodynamic limit as they are rounded off in finite systems. On the other hand, systems which are of infinite extension in two or more dimensions and which are unbounded in the remaining directions, show interesting dimensional crossovers as the transition is approached.[2] Then, it is important to extend theoretical approaches to understand finite systems with limiting surfaces.

Phenomenological finite-size scaling methods are widely used to extrapolate computer data to the thermo-dynamic limit,[2] but there are many aspects of finite size scaling which remain to be described by fundamental theories such as renormalization group (RG) methods. Such a fundamental theory becomes more important as interest extends to the study of particular finite systems with interacting boundaries. This is because universality classes of finite systems are more restricted than those of unbounded systems. The finite systems are character-ized not only by the dimensions of the embedding space and of the order parameter but also by the geometry of the system and the boundary conditions (BC) for the order parameter on the limiting surfaces.[2]

Here I discuss the application of field theoretic RG techniques to study critical phenomena in the presence of boundaries. The systems may be finite (or semi-infinite) along one (or several) of their dimensions, but they are of infinite extent in the remaining directions. Examples include systems which are finite in all directions, such as a (hyper) cube of size L, and systems which are of infinte size in $d' = d - 1$ dimensions but are either of finite thickness L along the remaining direction (e.g. a d-dimensional layered geometry) or of semi-infinite extension, etc. The presence of geometrical restrictions on the domain of systems also requires the introduction of BC (periodic, anti-periodic, free surfaces) for the order parameter on the surfaces.

Critical systems with boundaries or interfaces display a very rich set of phenomena because the (totally or partially) finite and semi-infinite cases contain several competing lengths and hence have interesting crossover behaviors as these length scales vary with respect to each other. These additional lengths are either associated with the finite size of the system in one or more of their dimensions or to the boundary conditions on the order parameter ϕ.[1,2]

Consider, for example, a semi-infinite critical system with a scalar order parameter which satisfies either the Neumann or the Dirichlet BC at the surface. These two cases belong to different universality classes called the special and ordinary transitions, respectively.[1] A surface interaction parameter c is usually introduced as $(1/\phi)(\partial\phi/\partial n)|_{\partial\Omega} = c$ where $(\partial\phi/\partial n)$ stands for the normal derivative of ϕ at the limiting surface $\partial\Omega$.[1] Then, as c ranges from zero to infinity the system crosses over from the special to the ordinary transition. These transitions are characterized, among other things, by different surface critical exponents.[1] On the other hand, systems that are bounded in one direction but of infinite extent in the remaining ones show a very interesting dimensional crossover as follows: In the critical domain, but away from the critical point, the behavior is dominated by the non-trivial 3d bulk fixed point, while as the transition is approached the 2d fixed point controls the physics.[2]

Section 2 shows that Feynman rules of field theories in configuration space are independent of geometry and boundary conditions, so they are *identical* to the well-known rules for unbounded systems. Geometrical constraints and boundary conditions are implemented through the explicit form of the zeroth order two point correlation function. Semi-infinite critical behavior is briefly discussed in Section 3 where we introduce a model of two coupled semi-infinite critical systems which possess a very rich physics. Section 4 considers the renormalization of field theories with boundaries and discuss a crossover renormalization group approach that is very well suited to describe interesting multiple crossovers present in these field theories with boundaries.

Section 5 deals with other interesting geometries. We begin by briefly discussing curved surfaces and edges, and then pass on to layered geometries with various boundary conditions (such as periodic, anti-periodic, Dirichlet and Neumann), and to cubic and cylindrical geometries. An important conclusion is that the usual ε-expansion technique *can be utilized* to study any geometry and boundary condition as long as the smallest finite system size is not much smaller than the bulk correlation length of the system.

Field theoretic methods can also be utilized to study the statistics of long polymer chains in solution near (liquid-liquid, liquid-solid) interfaces or in confined domains (such as a polymer chain in a cylindrical pore). This is the theme of Section 6. Finally, in Section 7 we present some analogies between the statistical mechanical problems of the preceding sections and related problems in quantum field theories.

2. Indefinite Integral Representation of Field Theories

Functional integrals provide an indefinite integral representation of the differential equations of a field theory. However, this representation *does not* contain a complete specification of the boundary conditions. Hence, the same functional integral representation of a field theory applies for various boundary conditions.[3]

Consider, for example, an $O(N)$ N-vector scalar ϕ^4 field theory in $d = 4 - \varepsilon$ dimensions in a region of the space Ω with a $(d-1)$ dimensional boundary $\partial\Omega$. The partition function $Z[J]$ is a functional of the external source J given by

$$Z[J] = \int D[\phi]\exp\left[-F\{\phi\} - \int_\Omega d^d x J(x)\phi(x)\right], \tag{2.1}$$

$$F\{\phi\} = \int_\Omega d^d x\left\{\frac{1}{2}[\nabla\phi(x)]^2 + \frac{1}{2}t_0[\phi(x)]^2 + \frac{u_0}{4!}[(\phi(x))^2]^2\right\}, \tag{2.1a}$$

where F is the free energy functional, $D[\phi]$ represents the sum over all configurations of the order parameter $\phi(x)$, x is a d-dimensional position vector inside the region Ω, $t_0 \propto T - T_c$, with T_c the (mean field) bulk critical temperature, and u_0 are the bare reduced temperature and coupling constant, respectively.

It is possible to formally integrate Eq. (2.1) over $\phi(x)$ to obtain

$$Z[J] = N^{-1}\exp\left\{-\int_\Omega d^d x \ u_0 \left[\frac{\delta}{\delta J(x)}\right]^2\right\}\exp\left\{\frac{1}{2}\int_\Omega d^d x \int_\Omega d^d x J(x) G^{(0)}(x, x') J(x)\right\}, \tag{2.2}$$

where N is a normalization constant such that $Z[J = 0] = 1$ and $G^{(0)}$ is the bare propagator (two-point correlation function) which is the solution to the usual Klein-Gordon wave equation

$$(-\nabla^2 + t_0) G^{(0)}(x, x') = \delta^{(d)}(x - x') , \tag{2.3}$$

Eq. (2.3) is satisfied in the region Ω and it must be supplemented with appropriate boundary conditions at $\partial\Omega$. Equivalently, $G^{(0)}(x, x')$ in (2.2) is *only* properly defined when boundary conditions are specified. The integral representation (2.2) of the ϕ^4 field theory is indefinite and applies to arbitrary boundary conditions which are implemented through the *properly chosen* propagator $G^{(0)}$. Coordinate space Feynman rules follow from (2.2), so they are independent of the explicit form of $G^{(0)}(x, x')$. Hence, the above discussion implies that *position space diagrammatic rules remain unchanged from those of an infinite volume theory, but that the appropriate zeroth–order propagator* $G^{(0)}(x, x')$ *must be utilized.* Chapter 14 of Ref. 4 contains expressions for the zeroth order two-point Green's function (in the context of the heat conduction problem) for a wealth of geometries and boundary conditions.

Translationally invariant systems have $G^{(0)}(|x, x'|)$ but, in general, the presence of interacting surfaces breaks this symmetry making $G^{(0)}(x, x') \neq G^{(0)}(|x - x'|)$. The n-point Green function also depends on *all n* coordinates rather than on n-1 coordinate differences as in full space. Diagrammatic expansions for unbounded systems are more conveniently performed in momentum space[5] where the translational invariance of the theory is reflected in momentum conservation conditions. The "most" convenient choice for finite systems depends on geometry and BC as discussed in Ref. 3. In the following sections we discuss various geometries and BC.

3. Critical Behavior at Surfaces

Semi-infinite critical systems have been studied by several workers using a variety of methods as described by Binder in his comprehensive review on the subject.[1] Renormalization group techniques have proven to be one of the most powerful theoretical techniques to study critical phenomena at surfaces. An excellent review by Diehl describes recent advances in this area.[6] Thus, the topics presented below sketch out very recent results which, in general, are not covered in either review. The interested reader may find the details in the reviews of Refs. 1 and 6 and in the original papers.

3.1. Semi-Infinite Geometry. Two Coupled Semi-Infinite Systems.

We begin with the usual Ginzburg-Landau free energy functional of (2.1a) in a semi-infinite geometry. Thus, the region Ω is the positive half-space $z > 0$ bounded by the $(d$-1)-dimensional flat surface $\partial\Omega$ at $z = 0$.

The position vector **x** of (2.1) is decomposed into its Cartesian components ρ and z with ρ a $(d-1)$-dimensional position vector parallel to the surface $\partial\Omega$ at $z = 0$.

Mean field theory predicts the appearance of four phase transitions depending on the values of the reduced temperature t_0 and the surface interaction parameter c_0 (introduced through the boundary conditions satisfied by $G^{(0)}$ at $\partial\Omega$ as discussed in Sec. 1).[1] These phases are depicted in Fig. 1a. For $c_0 \geq 0$ the system orders at the bulk critical temperature $t_0 = 0$. When c_0 is large, or more precisely when $c_0 \gg t_0^{1/2}$, the transition is called ordinary, while for small values of c_0, such that $c_0 \ll t_0^{1/2}$, it is known as the special transition. For $c_0 < 0$ the surface orders spontaneously at the surface transition, $(t_0 = |c_0|^{1/2})$ while the bulk remains disordered until $t_0 = 0$ (the extraordinary transition) where it also orders.

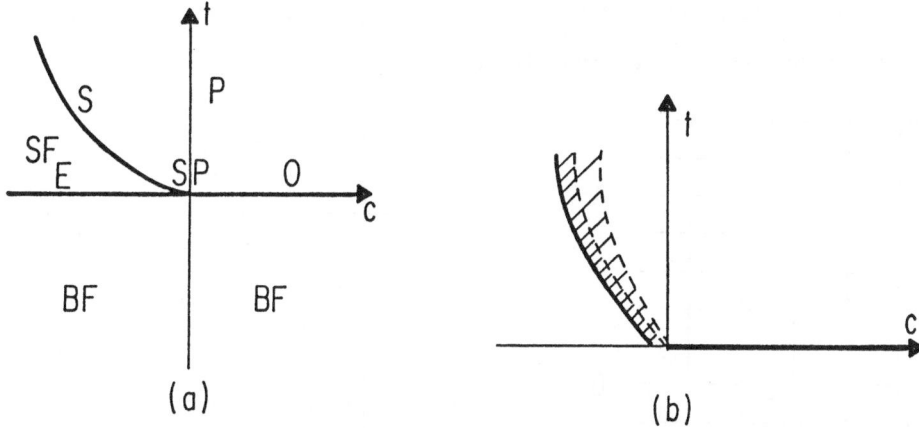

FIGURE 1

(a) Mean field predicts three phases depending on the values of the reduced temperature t and the surface interaction parameter c for semi-infinite systems: a paramagnetic (P) phase for $t > 0$ and $-t^{1/2} < c < \infty$, a bulk ferromagnetic (BF) phase for $t < 0$ and any c, and a surface ferromagnetic (SF) phase bounded by the P and BF phases. They are separated by the ordinary (O), surface (S) and extraordinary (E) lines. There is a tri-critical point at $t = c = 0$; this is the special (SP) transition. (b) Three regions can be distinguished in the paramagnetic phase. The physics in the unshaded region (the ordinary and the special transitions and the crossover between them) can be studied using usual ε-expansion techniques. These techniques cannot be employed in the dark shaded region (near the surface transition) where $d' = d-1$ physics emerges. Finally, there is an intermediate (gray) region of dimensional crossover where ε-expansions break down.

Recently, we[7] have considered a continuum theory of two coupled inequivalent semi-infinite systems based on suggestions in an alternative form by Diehl et al.[8] This theory was originally designed to understand the failure of the interesting conjecture of Bray and Moore[9] that the semi-infinite surface crossover exponent ϕ_s equals $1 - \nu$, where ν is the usual bulk correlation length exponent. Nevertheless, due to technical difficulties (such as the presence of "quasi-local" couplings), Diehl et al.[8] failed to construct an explicit realization of the theory, but their analysis of the problem enabled a realization to be devised by alternative means.[8]

Before discussing the continuum model for the two coupled systems, it is convenient to introduce a discrete version as illustrated in Fig. 2a. The model in Fig. 2a consists of two coupled semi-infinite spin systems on a d-dimensional hypercubic lattice. Neighboring spins in each of the semi-infinite systems A and B interact with bulk coupling constants J_b^A and J_b^B. Surface spins have different interaction constants $J_{\parallel}^A$ and $J_{\parallel}^B$, respectively. The two semi-infinite systems are coupled by perpendicular bonds with interaction strength $J_{\perp}$. When $J_{\perp} = 0$, the two systems decouple. For $J_{\perp} \to \infty$ the two surface layers "collapse" onto a layer with the nearest neighbor interaction $J_{\parallel} = J_{\parallel}^A + J_{\parallel}^B$ as shown in Fig. 2b.

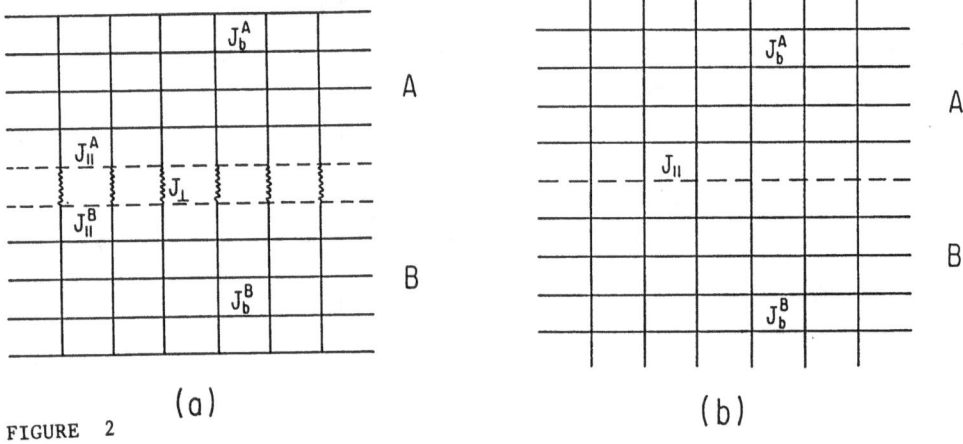

FIGURE 2

Figure 2a illustrates a discrete version of the model of two coupled semi-infinite systems of Section 3 whose symmetric limit ($J_{\parallel}^A = J_{\parallel}^B$ and $J_b^A = J_b^B$) is discussed in detail by Ref. 7. The model consists of two coupled semi-infinite spin systems on a d-dimensional hypercubic lattice. Neighboring spins in each of the semi-infinite systems A and B interact with bulk coupling constants J_b^A and J_b^B, respectively. Surface spins have interaction constants $J_{\parallel}^A$ and $J_{\parallel}^B$ that differ from the bulk ones. The two semi-infinite systems are coupled by perpendicular bonds with interactions $J_{\perp}$. When $J_{\perp} = 0$, the two systems decouple. For $J_{\perp} \to \infty$ the two surface layers "collapse" onto a single layer with a single surface parameter $J_{\parallel} = J_{\parallel}^A + J_{\parallel}^B$ as shown in Fig. 2b.

The full phase diagram of the model of Fig.2a is very rich. In addition to the (double-bulk and double-surface) paramagnetic phase, it exhibits four distinct broken symmetry phases as follows: a double-bulk, a single-bulk, a double-surface, and a single-surface ferromagnetic phase. An experimental realization of this model can be obtained by inserting a thin layer of a magnetic material X between two magnetic bulk materials A and B. If the two bulks are identical (and consequently $J_\parallel^A = J_\parallel^B$ and $J_b^A = J_b^B$), then the model reduces to that of an infinite critical system with a defect plane.

The continuum Ginzburg-Landau free energy of the model is again given by (2.1a), but now Ω contains both the positive and negative half-spaces, i.e., $\Omega = \{x = (\rho, z) \text{ with } z > 0 \text{ and } z < 0\}$ and $\partial \Omega = \{x = (\rho, z) \text{ with } z = 0_+ \text{ and } z = 0_-\}$. In general, the reduced temperatures (t_A, t_B) and coupling constants (u_A, u_B) of the two half-spaces can be different. To completely define the model, the boundary conditions for the (zeroth order in u_A and u_B) two-point correlation function $G^{(0)} \equiv G^{(0)}(|\rho - \rho'|, z, z')$ are specified on the surfaces at $z = 0_+$ and $z = 0_-$ by[7]

$$\frac{\partial G^{(0)}}{\partial z}\Big|_{0_+} = c_{A,0} G_+^{(0)} + \hat{c}_0 (G_+^{(0)} - G_-^{(0)}), \tag{3.1a}$$

$$-\frac{\partial G^{(0)}}{\partial z}\Big|_{0_-} = c_{B,0} G_-^{(0)} + \hat{c}'_0 (G_-^{(0)} - G_+^{(0)}), \tag{3.1b}$$

$$\hat{c}_0 = \hat{c}'_0, \tag{3.1c}$$

where the arguments of $G^{(0)}$ have been dropped, and $\partial G^{(0)}/\partial z|_{0_+}$ and $G_+^{(0)}$ stand for $\partial G^{(0)}(|\rho - \rho'|, z, z')/\partial z|_{z \to 0_+}$ and $G^{(0)}(|\rho - \rho'|, z \to 0_+, z')$, respectively. The bare surface interaction parameters are denoted by $c_{A,0}$, $c_{B,0}$ and $\hat{c}_0$. *The latter one couples the two semi-infinite systems.* Since the (zeroth order in u_A and u_B) two point Green's function satisfies the second order partial differential equation (2.3), boundary conditions are specified by given functions of $G_+^{(0)}$, $G_-^{(0)}$, $(\partial G^{(0)}/\partial z)_{0_+}$ and $(\partial G^{(0)}/\partial z)_{0_-}$ on the surfaces at $z = 0_+$ and $z = 0_-$.

The boundary conditions (3.1a) and (3.1b) are the most general linear and homogeneous ones relating $\partial G^{(0)}/\partial z|_{0_+}$, $(\partial G^{(0)}/\partial z)_{0_-}$, $G_+^{(0)}$ and $G_-^{(0)}$. Eq. (3.1c) is required to obtain a symmetric Green's function (in its arguments) i.e., $G^{(0)}(|\rho - \rho'|, z, z') = G^{(0)}(|\rho' - \rho|, z', z)$. The presence of surface magnetic fields h_1^A and h_1^B produces inhomogeneous contributions to (3.1a) and (3.1b). We take $h_1^A = h_1^B = 0$ for simplicity. Finally, the fact that we are considering renormalizable scalar field theories near four dimensions does not permit the use of nonlinear boundary conditions. (Nonlinear terms are associated with irrelevant surface operators near $d = 4$).

As discussed above the presence of surfaces breaks the translational invariance of the theory along the direction perpendicular to the surfaces. Of course, momentum is conserved in the direction parallel to the surfaces and, hence, $G^{(0)} \equiv G^{(0)}(|\rho - \rho'|, z, z')$. The function $G^{(0)}$ is, in general, discontinuous at $z(z') = 0$. (It only becomes continuous for $\hat{c}_0 = \infty$). This should not come as a surprise since a similar discontinuity is well known to occur in other contexts. For example, the presence of dielectric layers produces a jump in the potential, and the temperature is not continuous, in general, at the surface of separation of two media of different conductivities.

After some algebra and using equations (2.3) and (3.1), the zeroth order two-point correlation function $G^{(0)}(\mathbf{p}, z, z')$ with $\mathbf{p}$ the momentum variable conjugate to $\rho - \rho'$ is found to be of the form

$$G^{(0)}(\mathbf{p}, z, z') = G_{AA}^{(0)}\theta(z)\theta(z') + G_{BB}^{(0)}\theta(-z)\theta(-z') \tag{3.2}$$
$$+ G_{AB}^{(0)}\theta(z)\theta(-z') + G_{BA}^{(0)}\theta(-z)\theta(z') ,$$

The functions $G_{AA}^{(0)}$, $G_{AB}^{(0)}$, $G_{BA}^{(0)}$ and $G_{BB}^{(0)}$ are presented in Ref. 7 and some limiting cases are of interest. When $\hat{c}_0 = 0$, then $G_{AB}^{(0)} = G_{BA}^{(0)} = 0$, while $G_{AA}^{(0)}$ and $G_{BB}^{(0)}$ become identical to the two-point correlation functions of semi-infinite systems with surface interaction parameters $c_{A,0}$ and $c_{B,0}$, respectively.[1] When $\hat{c}_0 \neq 0$, the two semi-infinite regions are coupled. The $\hat{c}_0 \rightarrow \infty$ limit produces the single surface interaction model of Bray and Moore[9] with $c_0^{(BM)} = c_{A,0} + c_{B,0}$. As can be seen, the model of two coupled semi-infinite systems describes a very rich physical situation. Even the exactly solvable Gaussian theory with $u_{A,0} = u_{B,0} = 0$ is of interest and is far from trivial.

4. Renormalization of Field Theories with Boundaries

As stated in Section 2, Feynman rules of field theories in configuration space are independent of geometry and boundary conditions. These constraints are implemented through the explicit forms of the two-point functions so, for the problem of two coupled semi-infinite systems, the usual free propagator is replaced by the Fourier inverse of (3.2), and standard Feynman rules[5] are utilized to evaluate diagrams. Nevertheless, the breaking of translational invariance introduces novel features in these problems such as the presence of one-particle reducible primitively divergent diagrams as shown in Fig. 3 and as discussed in length by Ref. 6.

Renormalization of field theories in presence of interacting boundaries has been studied by Symanzik,[10] by Diehl and Dietrich,[11] and more recently by Diehl[6] and by us.[3] In addition to bulk renormalization constants Z_ϕ, Z_τ and Z_u which remain unchanged by the presence of surfaces (as do the β function and the fixed points $\{u^*\}$), it is necessary to introduce two additional renormalization functions Z_C and Z_1 required to renormalize surface interaction parameters and the fields on the surface. In the two coupled semi-infinite systems problem Z_C becomes a 2 x 2 non-symmetric real matrix.[7]

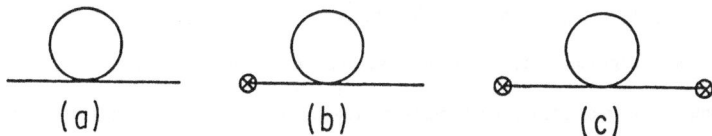

FIGURE 3

A new feature associated with the breaking of translational invariance is the existence of one-particle reducible primitively divergent diagrams. For example, the bare two-point functions G, G_1 and G_{11} with 0, 1 and 2 points on the surface of (a), (b) and (c), respectively, have different singularities, thus requiring different renormalization constants as discussed in Section 4.

With a semi-infinite geometry it is useful to work in a mixed momentum-configuration space representation. For example, we define the bare n-point connected Green's function $G_B^{(n)}$ $(p_i, z_i, c_0, t_0, u_0)$ [with c_0 the column vector $(c_0, \hat{c}_0)$ for the problem of coupled semi-infinite systems] as

$$G_B^{(n)}(\rho_i, z_i, c_0, t_0, u_0) = \left[\prod_{i=1}^{n} \int \frac{d^{d-1}p_i}{(2\pi)^{d-1}} \exp(-ip_i \cdot \rho_i)\right](2\pi)^{d-1}\delta^{d-1}(\sum_{i=1}^{n} p_i) \qquad (4.1)$$
$$G_B^{(n)}(p_i, z_i, c_0, t_0, u_0)$$

where $G_B^{(n)}(\rho_i, z_i, c_0, t_0, u_0)$ is the bare n-point function in configuration space, and the δ function reflects the fact that momentum is conserved in the direction parallel to the surfaces. The renormalized Green's function $G_R^{(n)}(p_i, z_i, c, t, u, \kappa)$ is then given by

$$G_R^{(n)}(p_i, z_i, c, t, u, \kappa) = Z_\phi^{-n/2}[Z_1^{(n)}]^{-1/2}G_B^{(n)}(p_i, z_i, Z_c c, Z_t t, S_d^{-1}\kappa^2 Z_u u), \qquad (4.2)$$

where κ is a parameter having dimensions of (temperature)$^{1/2}$ used to define a dimensionless coupling constant u and S_d is the area of a sphere of unit radius. Minimal subtraction dimensional regularization is the most widely used technique to renormalize these field theories.[6] Minimal subtraction has $Z_1^{(n)}$ of the form

$$Z_1^{(n)} = [Z_1(u)]^m, \text{ if } z_i = 0, i = 1, 2, ..., m, \text{ and } z_i \neq 0, i = m + 1, ..., n. \qquad (4.3)$$

Although this renormalization procedure is very convenient to study the physics near a given fixed point [FP] such as the special FP, the ordinary FP, the bulk FP, etc.,[6] minimal subtraction techniques are not well suited to describe the rich crossovers of field theories with boundaries with two or more competing fixed points.[12]

We have recently proposed a crossover RG approach that is very convenient for studying critical phenomena with several competing lengths.[12] Amit and Goldschmidt[13] utilize mathematically similar techniques to fully describe the bicritical crossover. In our coupled systems problem the surface normalization constants are taken to depend on the extra lengths through the dimensionless combination κz, c/κ, and $\hat{c}/\kappa$. This dependence emerges naturally by imposing appropriate normalization conditions on the two- and four-point Green's functions of the theory. In contrast, minimal subtraction dimensional regularization has the normalization constants independent of these lengths but only dependent on u and ε.

Due to the explicit κ-dependence of the renormalization constants in addition to the usual implicit κ-dependence through the dimensionless coupling constant u, the renormalization group equations become more involved, but they now describes the full crossover between all fixed points. Consider, for example, a semi-infinite critical system near the special transition. We have evaluated to one-loop approximation[12] the full z-dependence of the surface susceptibility $\chi_{11}(z)$ describing the response of spins in a plane at a distance z away from the surface at $z = 0$ due to a magnetic field applied on the same plane.

At the non-trivial fixed point u^*, the crossover RG equation implies the following scaling form for the renormalized surface susceptibility $\chi_{R,11}$

$$\chi_{R,11}(z, t, u^*, \kappa) = \kappa^{-1 + 2\nu(1 - \eta)} t^{-\nu(1 - \eta)} g(x, y) F(x) , \tag{4.4}$$

where ν and η are the usual bulk exponents, $y = \kappa z$ and $x = \kappa z (t/\kappa^2)^{\nu}$. Ref. 12 presents the functions $g(x, y)$ and $F(x)$ to $O(\varepsilon)$, and here we only give some interesting limiting cases. We always consider the asymptotic limit $t \ll \kappa^2$, but the magnitude of κz remains at our disposal. Thus, $y = \kappa z$ is always larger than $x = (\kappa z)(t/\kappa^2)^{\nu}$ and three regimes exist.

The $x \to \infty$ limit gives[12]

$$g(x, y) \to \exp[\varepsilon x^{-1/2} \exp(-x)] , \tag{4.5a}$$
$$F(x) \to 1 + O[\exp(-x)] , \tag{4.5b}$$

where higher order corrections, varying as $x^{-1} \exp(-x)$, have been dropped. Eqs. (4.4) and (4.5) imply that bulk behavior is approached exponentially fast for $(\kappa z)(t/\kappa^2)^{\nu} \gg 1$. This is in agreement with some previous discussion,[6] although Ref. 12 explicitly evaluates the corrections to bulk behavior in $O(\varepsilon)$.

When $x \ll 1$, two regimes are possible according to the value of $y = \kappa z$. If $y \geq 1$, we obtain[12]

$$g(x, y) = C(y) x^{-[(N + 2)/(N + 8)]\varepsilon} , \tag{4.6a}$$
$$F(x) = 1 + O(x, x \ln x) , \tag{4.6b}$$

where $C(y)$ is a finite function of y. The form predicted by (4.4) and (4.6) is in accord with scaling assumptions and previous calculations using minimal subtraction.[6] We stress that this near surface behavior is a *direct consequence* of the *full* crossover renormalization group approach. There is *no* need to utilize operator product expansion techniques.[12] Instead, these techniques are only required in the usual minimal subtraction approach because the standard RG equation does not contain information about the near surface behavior.[6] Finally, as $y \to 0$ we find[12]

$$g(x, y) = \tilde{C}(y) \, (x/y)^{-[(N+2)/(N+8)]\varepsilon} ,$$ (4.7a)

$$F(x) = 1 + 0(x, x \ln x)$$ (4.7b)

$$\tilde{C}(y) = 1 - [(N+2)/(N+8)]\varepsilon \, y \, \ln(y/2) .$$ (4.7c)

Hence, as $\kappa z \to 0$, $\chi_{R,11}(z)$ reduces to $\chi_{R,11} = \chi_{R,11}(z=0)$ as expected physically and in contrast to the results of the minimal subtraction renormalization approach. It is interesting to note that $\chi_{R,11}(z)$ is a continuous function of z for $0 \le z < \infty$, *but* no derivates exist at $z = 0$. Thus, it cannot be Taylor expanded about $z = y = 0$.

We have also evaluated the surface susceptibilities $\{\chi_{R,11}\}$ for the symmetric version of the model of two-coupled semi-infinite critical system discussed in Sec. 3.[7] Since there are two surface magnetizations m_1^A and m_1^B on either side of the interface and since a surface external field $h_1^A (h_1^B)$ can be applied to the positive (negative) side of the interface, four surface susceptibilities $\{\chi_{11}\}$ can be defined. The symmetric model $(c_A = c_B = c, u_A = u_B = u)$ only has the two susceptibilities. These susceptibilties are calculated near the special, ordinary and bulk fixed points.[7] Our results are in agreement with the qualitative discussion of Eisenriegler and Burkhard[14] based upon plausible renormalization group flow properties of the model of Fig. 2a. In particular, we find that the two crossover exponents ϕ_s and ϕ_s' near the special fixed point are the usual semi-infinite crossover exponent ϕ_s and the special exponent γ_{11}, respectively. Both of these exponents are given in Ref. 6 to $O(\varepsilon^2)$. Near the bulk fixed point only the surface interaction parameter c is relevant, and its associated surface exponent ϕ_s equals $(1 - \nu)$ as argued by Bray and Moore.[9] Finally, from our work it can be inferred that the infinite and semi-infinite ordinary transitions belong to the same universality class as predicted by Ref. 14 since the two limiting cases $(c/\kappa) \to \infty$ with $(\hat{c}/\kappa) = 0$ and with $(\hat{c}/\kappa) \to \infty$ yield the same surface susceptibilities in leading order. (Of course, they have different corrections).

5. Other Geometries

5.1. Curved Surfaces and Edges.

Curvature effects in critical systems have barely been considered beyond mean field theory. Recently, Eisenriegler[15] has studied ϕ^4 field theory inside a $d = 4 - \varepsilon$ dimensional hypersphere near the ordinary transition to one-loop order, and Diehl has briefly discussed curvature effects in his recent review.[6]

Consider, for example, a critical system bounded by the exterior surface of a sphere of radius R close to the special transition (the outside problem). Then, as the ratio (ξ/R) ranges from zero to infinity, the behavior of the system crosses over from that of a semi-infinite system at the special transition to bulk behavior. We expect this crossover to be described using the crossover renormalization group approach with surface normalization constants Z_1 and Z_c that depend on κR in analogy with problems discussed in Sec. 4. More generally, for an arbitrary curved surface, we believe that the surface constants should depend on $\kappa\{R\}$, where $\{R\}$ stands for the local radius of curvature of the surface.

Cardy[16] has studied critical phenomena near an edge where there is an edge interaction parameter δ (analog to the surface interaction parameter c) that measures how interactions near the edge differ from those deep in the bulk. In addition to the bulk and surface renormalization constants, there are now two additional edge renormalization constants Z_δ and Z_E (analogs of Z_c and Z_1) that renormalize the edge interaction constant and the fields on the edge respectively. Cardy has considered the case $c = \infty$ and $\delta = \infty$ (analogous to the ordinary transition). He finds a *new angular dependent critical edge exponent* (even Gaussian edge exponents are angular dependent!).

5.2. Layered Geometries

Before our recent work[17] it was widely believed that ε-expansion techniques were inapplicable for treating finite size problems.[2] We considered[17] a layered geometry (infinite in $d' = d-1$ directions and of thickness L in the remaining one) with periodic boundary conditions (BC), and we employed analogies with mathematically similar problems in finite temperature field theories to illustrate the applicability of the ε-expansion methods away from the critical point. Since then, other geometries and BC have been studied[15,3] with well defined ε-expansions.

Our studies of field theories in confined regions have used a variety of BC. We show[3] that as long a $(L/\xi)>1$, where ξ is the correlation length, ε-expansion techniques can be utilized to describe corrections to bulk quantities due to the finite extent of the system. Similar general results were derived by us for the effects of interacting boundaries where L is a parameter associated with surface interactions (as briefly summarized in Fig. 1b). The theory is illustrated for the N-vector model in a layered geometry with periodic, anti-periodic, Dirichlet and Neumann BC where the correlation functions and susceptibilities are evaluated to $O(\varepsilon)$. Away from the critical point and when $(L/\xi)\to\infty$, we find that first order contributions to scaling functions due to finite size are exponentially small, proportional to $\exp(-L/\xi)$, for periodic and anti-periodic BC, while these corrections behave as (ξ/L) for free surfaces. This is in accordance with previous numerical calculations and results obtained from various models.[2]

As the scaling variable (L/ξ) approaches unity, we show that first order in ε corrections to scaling amplitudes become comparable with zeroth order terms. This marks the beginning of a dimensional crossover where ε expansion methods break down. The finite size scaling literature[2] usually states that dimensional crossover occurs when the *bulk* correlation length becomes comparable to the typical system size L. While this is demonstrated by us to be true for a layered geometry with periodic or Neumann BC, it does not hold for example, for a layered geometry with anti-periodic or Dirichlet BC for which the ε-expansions are well behaved even at the *bulk* critical temperature T_c.[3,15]

Close to the transition a region of dimensionally reduced physics emerges. Layered systems near the shifted critical temperature and semi-infinite geometries near the surface transition have $d' = d-1$. Of course, different geometries, such as an infinite cylinder, a cube, etc., give different d'. We discuss[18] two mechanism for producing dimensional reduction (the emergence of d'-dimensional physics out of an underlying d-dimensional system): a geometrical one (e.g., a layered geometry very close to the shifted critical point), and an interaction drive one (e.g., a semi-infinite system close to the surface transition). An L dependent d' dimensional effective free energy functional for the lowest mode of the order parameter (massless mode) is evaluated by integrating out the higher (heavy) modes. Our approach presents some conceptual difficulties that still remain to be understood to fully describe the dimensional crossover. Can the crossover renormalization group approach be applied to this problem? We are presently investigating this interesting possibility.

5.3. Cubes, Cylinders and Other Geometries. Dynamical Critical Phenomena and First Order Transitions

Our recent work[3] and that of Ref. 15 show that the usual ε-expansion techniques *can be applied* to study any geometry and boundary conditions as long as the bulk correlation length of the system is not much larger than the smallest dimension of the system. As the system approaches arbitrarily close to the critical (or pseudocritical) point, the ε-expansion break down. Related techniques to our effective free energy functional method have been proposed by Brézin and Zinn-Justin,[19] and by Rudnick et al.[20] to investigate the deep critical region for cubic and cylindrical geometries with periodic BC. Their approaches do not present the technical difficulties of ours as discussed above, since these authors only consider systems with no *true critical points*. Brézin and Zinn-Justin have also proposed a $2 + ε$ expansion to study finite size effects in critical phenomena below T_c.[19] Since then, several authors[21] have extended the methods of Refs. 19 and 20 to study finite size effects on dynamics and in first order transitions always for systems with no true critical points and with periodic boundary conditions.

6. The Statistics of Polymers in Various Geometries.

The study of conformational properties of long, flexible polymer chains near penetrable (liquid-liquid) or impenetrable (liquid-solid) interfaces or in various confined geometries (e.g., polymer chains in cylindrical or spherical pores) has a variety of important practical applications. These applications include cohesion, stabilization of colloidal particles, chromotography reinforcement and flocculation. Also, we note that finite-size effects are present in computer simulations of polymer systems. Simulations generally employ periodic boundary conditions to remove the surface interactions, but the finite size of the computer still affects the computed thermodynamic properties. Therefore, systematic extrapolation of the simulation data is required in order to describe properties of the infinite system. It is, therefore, of theoretical interest to understand how the thermodynamic limit is approached as the size of the system is increased.

The statistics of long flexible polymer chains with excluded volume in dilute solutions is well known to belong to the same universal class as that of the $O(N)$ ϕ^4 field theory with $N = 0$.[22] This holds not only for unbounded systems but also for systems with interacting interfaces and those in confined geometries. Thus, most of the results for critical systems discussed in the previous sections can be transcribed to corresponding polymer problems. We have used powerful field theoretic techniques to study the conformational properties of polymers near interacting impenetrable[23] and penetrable[24] interfaces and polymer chains in confined geometries[25] such as polymers between two parallel plates with various polymer-surface interactions on the limiting surfaces or polymers near the outside surface of a repulsive sphere. Some of the rich array of situations that can now be treated using renormalization group methods are illustrated in Fig. 4.

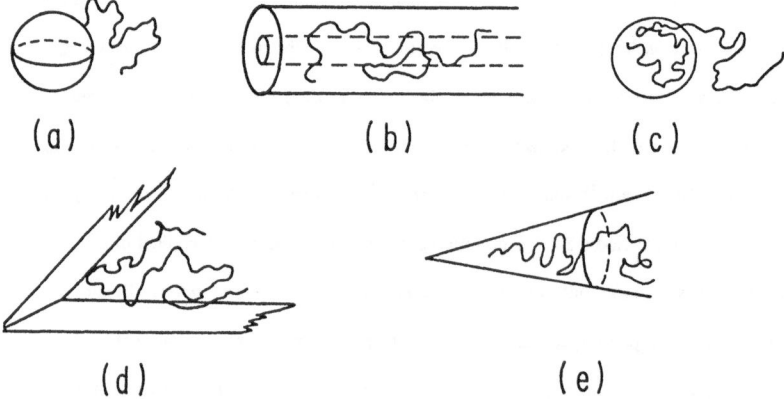

(a) (b) (c)

(d) (e)

These figures illustrate some interesting systems involving a single polymer chain with excluded volume and interacting boundaries in several geometries. These geometries can now be studied by employing the RG methods discussed in Section 6 [as long as the radius of gyration of the polymer chain is not longer than the smallest dimension of the system]:

(a) A polymer attached to a sphere with an interacting surface.

(b)   A polymer in the shell formed by two concentric cylinders.

(c)   A polymer near a sphere formed by two different solvents, e.g. oil and water. The quality of these two

        solvents is, in general, different. Furthermore, the interfacial region can be such that one side of the inter-

        face attracts the polymer whereas the other side repells it.

(d)   A polymer in an edge where the power law exponent for some property(ies) can depend on the edge

        angle.

(e)   A polymer in a cone.

7. Quantum Field Theories with Boundaries

It is well known that there are many analogies between statistical mechanics and quantum field theories (QFT) for unbounded systems.[5] For example, the Green's functions of the QFT's are the analogues of the correlation functions in statistical mechanics, and $Z[J]$ of (2.1) can be viewed as the generating functional of Euclidean self-interacting scalar QFT. Successive derivatives of $Z[J]$ respect to the external source J produce all the Green's functions of the theory. These analogies, of course, also hold when boundaries are present.

The Casimir effect, the attraction of two neutral and parallel plates in a vacuum environment, predicted and experimentally confirmed several years ago, is the earliest example of boundary effects in QFT.[26] An interesting example of the scalar "Casimir effect" in statistical mechanics, as discussed by Diehl,[6] is provided by fluctuation-induced force between two plates with a binary fluid mixture at its consolute point held in between.

Systems that are of infinite extent in two or more of their dimensions and finite in the remaining directions such as a layered geometry in $d=3$ dimensions, display $3d$ physics away from the shifted critical temperature (but inside the critical domain) but $d'=2$-dimensional physics in the deep critical region.[2] Dimensional reduction, the emergence of a quasi d' dimensional physics out of an underlying d dimensional system, is one of the main ingredients of the Kaluza-Klein theories.[27] In fact, Kaluza-Klein masses are the analog of *experimentally observed*[28] shifts in critical temperatures of finite size systems from those of the bulk.

We have used the analogy between finite size problems in a periodic layered geometry and similar problems in finite temperature field theories to demonstrate how ε-expansion techniques can be employed to study finite systems away from the shifted critical point as described in Sec. 5.2. At finite temperatures β^{-1} (where $\beta = (kT)^{-1}$, k is Boltzmann's constant and T is the absolute temperature) the causal boundary conditions of field theories in real time are replaced by periodic boundary conditions with period β in Euclidean time.[29] Thus, a finite temperature field theory is identical to one contained between two-parallel (hyper) plates with periodic boundary conditions. The (hyper) planes are perpendicular to the Euclidean time direction, and the periodicity is β.

ACKNOWLEDGEMENT

I am grateful to H.J. de Vega for his kind hospitality at Paris VI and to K. Binder, H.W. Diehl and E. Eisenriegler for useful discussions. This research is supported, in part, by NSF grant DMR 83-18560.

REFERENCES

1. For a review of surface effects near criticality see K. Binder, in *Critical Behavior at Surfaces*, *Phase Transitions* and *Critical Phenomena*, Vol. VIII, C. Domb and J.L. Lebowitz, eds. (Academic Press, N.Y., 1983).

2. For a review on finite size effects in critical phenomena see M.N. Barber, in *Finite Size Scaling*, *Phase Transitions* and *Critical Phenomena*, Vol. VIII, C. Domb and J.L. Lebowitz, eds. (Academic Press, N.Y., 1983).

3. A.M. Nemirovsky and K.F. Freed, Nucl. Phys. **B270**, [FS16], 423 (1986).

4. H.S. Carslaw and J.C. Jager, *Conduction of Heat in Solids* (Clarendon, Oxford University, Oxford, 1959).

5. D.J. Amit, in *Field Theory*, *the Renormalization Group* and *Critical Phenomena* (World Scientific, Singapore, 1984).

6. For a recent review on field theoretic techniques to critical phenomena near surfaces see H.W. Diehl, in *Field Theoretic Approach Critical Behavior at Surfaces*, *Phase Transitions* and *Critical Phenomena*, Vol. X, C. Domb and J.L. Lebowitz, eds., in press.

7. A.M. Nemirovsky, Z.-G. Wang and K.F. Freed, Phys. Rev. B, in press.

8. H.W. Diehl, S. Dietrich and E. Eisenriegler, Phys. Rev. **B27**, 2937 (1983).

9. A.J. Bray and M.A. Moore, J. Phys. **A10**, 1927 (1977).

10. K. Symanzik, Nucl. Phys. **B190**, [FS3], 1 (1981).

11. H.W. Diehl and S. Dietrich, Phys. Lett. **80A**, 408 (1980); Z. Phys. **B43**, 315 (1981); Z. Phys. **B50**, 117 (1983).

12. A.M. Nemirovsky, Z.-G. Wang and K.F. Freed, submitted to Phys. Rev.B.

13. D.J. Amit and Y. Goldschmidt, Ann. Phys. **114**, 356 (1978).

14. T.W. Burkhardt and E. Eisenriegler, Phys. Rev. **B24**, 1236 (1981).

15. E. Eisenriegler, Z. Phys. **B61**, 299 (1985).

16. J.L. Cardy, J. Phys. **A16**, 3617 (1983).

17. A.M. Nemirovsky and K.F. Freed, J. Phys. **A18**, L319 (1985).

18. A.M. Nemirovsky and K.F. Freed, J. Phys. **A19**, 591 (1986); ibid. **A18**, 3275 (1985).

19. E. Brézin and J. Zinn-Justin, Nucl. Phys. **B257** [FS14] 867 (1985).

20. J. Rudnick, H. Guo and D. Jasnow, J. Stat. Phys. **41**, 353 (1985).

21. Some authors are beginning to study finite size effects in first order transitions and in dynamics. See, G.G. Cabrera, R. Jullien, E. Brézin and J. Zinn-Justin, J. Physique **47**, 1305 (1986); J.C. Niel and J. Zinn-Justin, preprint; Y. Goldschmidt, preprint; H.W. Diehl, preprint.

22. See, for example, P.-G de Gennes, *Scaling Concepts in Polymer Physics* (Cornell University, Ithaca, 1979) and references therein.

23. A.M. Nemirovsky and K.F. Freed, J. Chem. Phys. **83**, 4166 (1985).

24. Z.-G. Wang, A.M. Nemirovsky and K.F. Freed, J. Chem. Phys. **85**, 3068 (1986).

25. Z.-G. Wang, A.M. Nemirovsky and K.F. Freed, submitted to J. Chem. Phys. J.F. Douglas, A.M. Nemirovsky and K.F. Freed, Macromolecules, **19**, 2041 (1986).

26. H.B.G. Casimir, Proc. Kon. Ned. Akad. Wetenschap., **B51**, 793 (1948). Experimental evidence is discussed by M.J. Sparnaay, Physica, **24**, 751 (1958).

27. For a description of Kaluza-Klein theories, see, for example, E. Witten, Nucl. Phys. **B186**, 412 (1981); A. Salam and J. Strathdee, Ann. of Phys. **141**, 316 (1982).

28. B.A. Scheibner, M.R. Meadows, R.C. Mockler and W.J. O'Sullivan, Phys. Rev. Lett. **43**, 590 (1979); M.R. Meadows, B.A. Scheibner, R.C. Mockler and W.J. O'Sullivan, Phys. Rev. Lett. **43**, 592 (1979); F.M. Gasparini, T. Chen and B. Bhattacharyya, Phys. Rev. **B23**, 5795 (1981) and references therein.

29. C.W. Bernard, Phys. Rev. **D9**, 3312 (1974); L. Dolan and R. Jackiw, Phys. Rev. **D9**, 3320 (1974); S. Weinberg, Phys. Rev. **D9**, 3357 (1974).

Lecture Notes in Physics

Astronomy and Astrophysics Library

Series Editors: M. O. Harwit,
R. Kippenhahn, J.-P. Zahn

K. Rohlfs

Tools of Radio Astronomy

1986. 127 figures. XII, 319 pages. ISBN 3-540-16188-0

The first part of this book gives a complete introduction to the instrumentation and techniques needed to do radio astronomical research. After a thorough survey of electromagnetic wave propagation and polarization, the text focusses on the antenna theory of both filled aperture antennas and of interferometers and aperture synthesis telescopes. The theory and design of receivers is thoroughly discussed covering front-ends as well as the various kinds of back-ends in use. Great care has been taken to give a representation useful for physicists not yet familiar with the technicalities of microwave physics.
Radiation mechanisms relevant to radio astronomy are subjects of the second part of the book. Chapters on thermal bremsstrahlung and synchrotron radiation responsible for the continuous spectrum are followed by an extensive treatment of line radiation. Discussions of the 21 cm-line of hydrogen, the recombination lines and the emission from interstellar molecules expose those tools of radio astronomy which allow the derivation of the physical parameters of the interstellar medium.
Combining for the first time material from widely separated fields in one volume, this well-illustrated textbook will be useful for students and teachers in astronomy as well as for the active researcher.

H. Scheffler, H. Elsässer

Physics of the Galaxy and Interstellar Matter

Translated from the German by A. H. Armstrong

1987. 210 figures. Approx. 450 pages. Hard cover.
ISBN 3-540-17314-5. Soft cover. ISBN 3-540-17315-3

This book is based on the authors' long-standing experience in teaching astronomy courses. It presents in a modern and complete way our present picture of the physics of the Milky Way system. The first part of the book deals in comprehensible terms with topics of more empirical character, such as the positions and motions of stars, the structure and kinematics of the stellar system and interstellar phenomena. The more advanced second part is devoted to the interpretation of observational results, i. e. to the physics of interstellar gas and dust, to stellar dynamics, to the theory of spiral structures and to the dynamics of interstellar gas.
Tailored for students and lecturers in related courses in astronomy, the book should be equally interesting for researchers working in other fields of physics and astronomy and, in particular, for the educated amateur astronomer.

Springer-Verlag
Berlin Heidelberg New York
London Paris Tokyo

Springer